T0184654

Knowledge Communities in Europe

Bertold Schweitzer · Thomas Sukopp
Editors

Knowledge Communities in Europe

Exchange, Integration and Its Limits

 Springer VS

Editors
Bertold Schweitzer
Dundee, Scotland

Thomas Sukopp
Siegen, Germany

ISBN 978-3-658-18851-1   ISBN 978-3-658-18852-8   (eBook)
https://doi.org/10.1007/978-3-658-18852-8

Library of Congress Control Number: 2017963268

Springer VS

Verantwortlich im Verlag: Frank Schindler

Printed on acid-free paper

This Springer VS imprint is published by Springer Nature
The registered company is Springer Fachmedien Wiesbaden GmbH
The registered company address is: Abraham-Lincoln-Str. 46, 65189 Wiesbaden, Germany

Table of Contents

Introduction

1

Bertold Schweitzer and Thomas Sukopp

1.1 Knowledge Communities and Political Integration: Two Complementary Perspectives

This volume attempts to explore and combine two fresh perspectives on the relation of science and politics: From the perspective of philosophy, history, and sociology of science, it seems worth investigating how scientific and other knowledge communities interact with their wider social and political surroundings, both in a national and an international context. From the perspective of political science, it appears fruitful to ask how existing inter- or transnational communities contribute to the strengthening of transnational ties, and to facilitate political cooperation and integration. Bringing together these two perspectives, we ask: How are knowledge communities in science and other fields formed, how do they function, and how do they interact with their wider environment?

These broad-ranging questions at the intersection of the social sciences and the philosophy and history of science have motivated the project "Knowledge Communities in Europe", the results of which are presented in this volume. More specifically, the project has been focusing on the question of how *transnational* knowledge communities are formed, what roles they play, and how they interact with processes of social and political *integration*.

1.2 Integration and the Role of Elites

One central motivation for this project stems from the idea that it may not only be the actions of *political* elites that promote social and political integration, but that other influential groups engaged in transborder projects, such as, for example, scientists, may play a significant role in integration processes, too.

1

This idea is an extension of one of the central tenets of the neofunctionalist school in political science, which has been claiming that political integration takes place when political actors shift their loyalties to a new level. The general patterns of how elites are involved in facilitating political integration have been described in *The uniting of Europe* by Ernst B. Haas:

> Political integration is the process whereby political actors in several distinct national settings are persuaded to shift their loyalties, expectations and political activities to a new center, whose institutions possess or demand jurisdiction over pre-existing national states. The end result is a new political community, superimposed over the pre-existing ones. (Haas 1958, 16)

Extending this idea, we introduce three assumptions. First, we contend that *political* "loyalties, expectations and [...] activities" are influenced to a significant part by *social* institutions—whether formal institutions, such as communities and organizations, or informal ones, such as norms, loyalties, expectations, and identities. Second, we accept that social institutions that tend to promote transnational cooperation in general are likely to facilitate political integration as well. Third, we take scientists and scientific communities to be among those groups who are, in general, either intrinsically inclined or can be easily persuaded to value and engage in transnational cooperation.

Based on these assumptions, we surmise that actors of an intrinsically strong transnational persuasion—as may be suspected of scientists and other members of knowledge communities—are likely to contribute, sometimes significantly, to transnational social and political cooperation or integration in certain ways.

This, of course should not imply that our understanding of political integration is restricted to a Haasian account. Furthermore, in what follows, the term "integration" will be understood in a way that is not limited exclusively to political integration. In fact, for most processes and developments in the sciences in general, and in biology in particular, *social integration* will turn out to be the most relevant phenomenon. Hence, integration is understood as consisting of multifaceted processes of socialization, knowledge acquisition and the establishment of discourses in which certain groups such as scientific and other elites play a central role.

The project's emphasis on the interrelation between knowledge communities and political integration in turn suggests focusing on examples from Europe and the European Union (in the following: EU), as the latter can be seen as the prime example of far-reaching political integration.

1.3 Three Key Questions

Based on these premises, this collection assembles contributions from history, philosophy, political science, and sociology, each of which focuses on one or more of the three interrelated questions:

1. How are transnational scientific communities formed, and how do they function?
2. How do transnational scientific communities affect transnational integration in general?
3. How does transnational integration in turn influence transnational scientific communities?

The first of these questions has been analyzed rather extensively from a number of different disciplinary perspectives—though not always exhaustively or without any remaining controversies. Hence, the following section will provide an overview of the different attempts at characterizing the notion of a "scientific community". By contrast, relatively little attention seems to have been paid to the second and the third question—which form the centrepiece of this project's approach.

Each of the contributions to this collection tries to answer one or more of the three key questions: While the chapters by Sebastian Nähr (on social epistemology) and by Thomas Sukopp (on scientific communities in chemistry) focus more on the role of communities in defining and establishing knowledge claims, and on the dynamics of communities in science, the chapter by Bertold Schweitzer explores the role of scientific communities in initiating or promoting increased transnational cooperation and integration. Finally, the chapters by Julie Patarin-Jossec, Patricia Bauer, and Stephen Rozée examine the mutual interaction of scientific and other knowledge communities with structures created by political integration, and how each of these exert influence on shaping the other.

1.4 Scientific Communities: Features and Functioning

The first complex of questions asks how scientific communities—including transnational ones—can be characterized, how they are formed, and how they function. This includes developing an understanding of the complex processes of knowledge acquisition, knowledge transformation and knowledge dissemination—both as individual and as social processes taking place in scientific and other knowledge communities.

A wealth of detailed work has been done in various fields, from disciplinary as well as from inter- and transdisciplinary perspectives. Sociology of science, history of science, philosophy of science, science and technology studies, history of intellectual ideas, epistemology, sociology and political science have been engaging with these questions to foster a better understanding of how knowledge is created, evaluated, and disseminated. Certain areas of research in fact turned out to be especially fruitful for the project and are in part developed in this volume:

Epistemic networks or epistemic communities, beyond constituting a fascinating subject in philosophy and sociology of science, traditionally have played an important role in theories of social and political integration. In philosophy and sociology of science, epistemic networks or communities have been variously described as being centred on shared problem perceptions and epistemological positions, as intentional communities, and as socio-psychological entities that create and justify knowledge. The functioning of scientific communities and epistemic networks relies on a shared language or languages—both natural languages, such as Latin, and later English, French, German and others as well as technical languages and vocabularies—, and on various mediums, such as personal visits, lectures, meetings and congresses, museums, laboratories, manuscripts, correspondence, publications (journals, books), pictures, maps, and objects.

As to the characterization of scientific, epistemic, or knowledge communities, at least three approaches may be distinguished, originating in the philosophy of science, the history and sociology of science, and political science. The next sections will introduce these approaches in turn.

1.4.1 Knowledge Communities in Philosophy of Science

In the areas of philosophy of science that focus on the structure and justifiability of scientific theories, scientific communities and social aspects in general traditionally played a small, if any, role. This approach, hence, is aptly described as the *logic*, as opposed to the psychology or sociology, of science (see, e. g., Popper 2002, 7, in which he urges philosophers to "separate the psychological from the logical and methodological aspects"). A stronger focus on *social* phenomena emerged only in the 1930s when Ludwik Fleck, in *Genesis and development of a scientific fact* (1981), introduced the notion of a "thought collective" as a sociological group with a common style of thinking. Much more influentially, Thomas S. Kuhn, in *The structure of scientific revolution* (1962), developed the concept of a "scientific community" consisting of scientists sharing one particular paradigm, or research programme. Kuhn used this concept to explain the processes of theory change by referring to social processes in which different communities share different scientific approaches, background assumptions, and world views, and used this to explain how new theories are formed, how they eventually become accepted in the scientific community, but also why scientific change is by no means always a smooth process, and how different and contradicting views can persist in science for considerable amounts of time. Though Kuhn's account has had to face manifold criticism (from, for example, Lakatos, Popper, Feyerabend, and postmodernist theorists), its principles continue to be useful for explaining numerous historical developments, in science and beyond.

1.4.2 Knowledge Communities in History and Sociology of Science

The term "scientific community" of course is not necessarily limited to describing groups of scientists sharing one particular research programme. For history and sociology of science, the focus on scientific communities of all kinds and sizes would indeed seem to be a natural one. Historians of science, however, have expressed their disappointment that "[t]he history of scientific institutions and communities is marked by a historiographical paradox: A large number of scholars—indeed most—routinely express their interest in the subject, yet relatively few pursue it in a systematic and concerted manner." (Cahan 2003a, 291) From the perspective of the history and sociology of science, a suitable definition of scientific communities emphasizes not only shared aims and interests, and a certain durability, but also, most importantly, shared values, and a shared identity:

> [...] scientific institutions and communities may be characterized as consisting of populations of individuals who share similar cognitive interests and values that serve to provide them with a collective social identity and to advance individual scientific careers and group needs. Such populations are naturally composed of individual scientists and their variegated associates, yet they only become institutions and communities when those individuals—perhaps only few in number—act in concert over an extended period of time and perceive themselves as bound together in some particular professional manner. When such individuals hold more or less similar cognitive interests and values and act more or less together as a group, we often speak of a "community," though scholars sometimes use the related (and usually undefined) notions of "discipline" or "school." (Cahan 2003a, 293)

In this view, the first appearance of such scientific communities in European science is associated with developments in the nineteenth century: "I argue that there was no identifiable scientific community before the early nineteenth century" (Cahan

2003b, 11). Only "by the final third of the nineteenth century, [...] new institutions, such as specialized societies and institutes, were created, and the notion of a 'scientific community' appeared." (Cahan 2003b, 4) He contends, however, that what he labels "protocommunities" had emerged in science by the late Enlightenment (Cahan 2003a, 295 f.).

Cahan discusses the manifold manifestations of institutions and communities, and their various and variable forms and functions: "[...] in principle a scientific institution or community may be an academic or nonacademic unit within a state or private structure; a physical building housing laboratories, classrooms, and the like; or, more abstractly, a set of social and professional relations among individuals." (Cahan 2003a, 293) This includes formal organizations, such as institutes, research laboratories, university departments, societies, associations, academic journals, as well as informal communities and institutions. The latter "include [...] the informal groupings of scientists ('networks,' 'invisible colleges,' and, again, 'schools') that, for instance, facilitate the transfer of scientific knowledge, skills, and values from individual to individual and generation to generation" (Cahan 2003a, 293).

The most important aspect here, however, is the collective identity of scientific communities: "[h]owever diverse such institutions and communities may be, to one degree or another they all share a collective identity" (Cahan 2003a, 293), combined with the historial judgement, also forming "... one of this essay's major claims: namely, that what are usually seen as discrete, individualized institutions and communities came during the nineteenth century to constitute an imagined yet real scientific community" (Cahan 2003a, 295). Cahan links these ideas with Benedict Anderson's reflections on the origin of nationalism in which the nation is defined as "an imagined political community [...] imagined as both inherently limited and sovereign" (Anderson 2006, 6). In scientific communities, too, "most members (scientists) would not have known most other members personally, yet they imagined themselves as belonging to one and the same community" (Cahan 2003a, 325). Finally, Cahan discusses three levels, of disciplinary, of national, and of international scientific communities in the nineteenth century: "The concept of a scientific community becomes still more abstract when applied on the international plane, yet such a community had also begun to manifest itself by midcentury", and "scientists [...] belonged to an international social unit" (Cahan 2003a, 328). It is these international scientific communities the present volume will be focusing on.

1.4.3   Knowledge Communities in Political Science

In political science, and with a stronger focus on the role of experts—frequently from science or related fields—in political processes, the idea of "epistemic communities" was made prominent by Burkart Holzner, in *Reality Construction in Society* (1968), and by Ernst Haas, Mary Williams, and Don Babai, in *Scientists and World Order* (1977), who characterised them as "networks of knowledge-based experts". More recently, in a special issue of *International Organization*, edited by Peter M. Haas (1992), "epistemic communities" are described as being equipped with recognized expertise and competence in a particular domain and an authoritative claim to policy-relevant knowledge within that domain or issue-area, including the following four-step definition:

> Although an epistemic community may consist of professionals from a variety of disciplines and backgrounds, they have (1) a shared set of normative and principled beliefs, which provide a value-based rationale for the social action of community members; (2) shared causal beliefs, which are derived from their analysis of practices leading or contributing to a central set of problems in their domain and which then serve as the basis for elucidating the multiple linkages between possible policy actions and desired outcomes; (3) shared notions of validity—that is, intersubjective, internally-defined criteria for weighing and validating knowledge in the domain of their expertise; and (4) a common policy enterprise—that is, a set of common practices associated with a set of problems to which their professional competence is directed, presumably out of the conviction that human welfare will be enhanced as a consequence. (Haas 1992, 3)

Here, the distinctive features of an "epistemic community", as opposed to, for example, a social movement or a political pressure group, are the second and third element: shared beliefs on the causal structure of the world, and an accepted methodology for establishing and evaluating specific causal claims. Though it has been disputed whether this amounts to an overly science-oriented view of epistemic communities (see, e.g., Toke 1999; Dunlop 2000), what seems clear is that the sciences themselves, whether pure or applied, certainly are seen as prime examples of epistemic communities.

1.5   Transnational Scientific Communities and Political Integration

The second complex of questions asks how transnational scientific communities affect transnational integration. We hypothesize that transnational scientific communities do in fact contribute positively to transnational integration in a number of ways: In general, they do so by promoting the values of science and, in particular, the value of transnational cooperation in science.

More specifically, joint *political* action introduced to facilitate transnational scientific cooperation can facilitate further political integration. In particular, some of the spillover effects postulated by the neofunctionalist school can be expected; for example, the exchange of scientists presupposes their free movement, and the collection and distribution of research funding requires overseeing bodies.

To answer the question of how transnational scientific communities affect transnational integration, we propose a number of heuristically fruitful, but nevertheless controversial basic assumptions. Each of these assumptions would seem to require further testing (a task which is, however, beyond the scope of this volume): First, we tend to assume that, in general, scientific elites are blind with regard to nation and ideology. However, it would seem to be rather naive to believe that nationalism is always absent from science, or that science is completely detached from ideologies or world views. Hence the default assumption of scientists usually favouring cooperation or integration is a useful one but still merits a critical discussion.

Another controversial assumption is that scientific communities establish and disseminate knowledge exclusively by rational discussion. But even if scientists, intellectuals, scholars, or experts in their fields are mainly driven by rationality, the formation of European knowledge communities is influenced in many respects by political processes and actors. This might seems rather trivial, but a detailed analysis on how scientific rationality and (decision) logic of, for example, EU research policy may shed light on the question of how policy and science mutually foster or impede integration processes (see the contributions of Bauer, Patarin-Jossec and Rozée in this volume).

1.6   Transnational Political Integration and Scientific Communities

The third complex of questions asks, in turn, how transnational scientific communities are influenced by transnational integration and the organizational structures created along with it. First, one manifest aim of transnational organizations promoting political integration is to reach out to citizens, trying to instil a sense of "cohesion" and "identity", and scientific projects have been used for this purpose alongside political and economic arguments. One example from the Director General of the European Space Agency claims: "Space can contribute to European cohesion and identity, reaching citizens across all countries" (Commission of the European Communities 2007). Likewise, capable administrative and logistic structures, high-prestige international research institutions, and a vision of Europe as capable of competing globally are other examples where science helps satisfying the needs of political integration processes and the actors invested in them.

Second, a number of research areas are highly dependent on formalized international cooperation, for example in elementary particle physics, or in space research. Here we can expect political systems to influence both the organization and, in part, the focus of the research being carried out.

Third, the EU's research and innovation policy exhibits tendencies towards influencing and shaping the character of scientific activities by the way science is funded and administered. These tendencies include that the definition of what is regarded as excellence in scientific research is in part becoming a political issue, and that sciences are restricted by basic conditions that are formed, transformed and regulated by policy. What is controversial is how independent research under the maxim of freedom of science can be upheld and maintained if the current vision of European research depends to a large extent on the postulated interdependency of research, innovation, technology, and its relevance for a globally competitive marked-based economy (for these and other aspects, see the contributions by Bauer on the European Research Area, by Patarin-Jossec on the European Space Agency, and by Schweitzer in the section on the European Molecular Biology Organization).

1.7   Summaries of the Contributions to this Volume

The general methodological approach used by the authors contributing to this volume may best be characterized as an integrated transdisciplinary approach that combines and evolves methods and results of general philosophy of science, philosophy of chemistry, philosophy of biology, history of science, social studies of science, and sociology of science, and political science.

In his contribution "Discoveries of Oxygen and the 'Chemical Revolution' in the Context of European Scientific Networks", Thomas Sukopp offers an analysis of the complex processes contributing to the formation of modern chemistry between the 1770s and the 1790s. Styles and modes of communication within frameworks of collaborative thinking in research groups are one key element to understand if and how Lavoisier can or cannot be regarded as the centre of the so called "chemical revolution". Lavoisier's "revolutionary" idea was to understand combustion as the combination of substances with the gas oxygen. The main rivalling paradigms, phlogiston theories vs. oxygen theories, can only be adequately understood if we consider them as articulated transnationally via the exchange of letters, the publication of books, and in personal meetings. Figures like Scheele and his German-Swedish scientific network were no outsiders in this, and neither were renowned scientists such as Cavendish, Priestley and others who remained adherents of the phlogiston theory. Core elements of Lavoisier's success have been analysed extensively by standard approaches of philosophy of science. However, most of these attempts, especially traditional reconstructions of scientific revolutions, such as plain interpretations of Kuhn, fail to convince. In favour of Lavoisier's merits as one main figure of the chemical revolution, Sukopp argues that Lavoisier's revolutionary stance combines at least two aspects: First, a rationalist view on chemistry as a science, and second, his engagement in and understanding of a research group in a rather modern way, i.e., a genuinely collaborative way of conceptualizing chemistry and carrying out chemical experiments.

In the following chapter "From interaction to integration? Transnational scientific communities in biology", Bertold Schweitzer discusses how networks, clubs, and international scientific organisations have shaped cooperation in the biological sciences and how these processes have influenced political integration. His analysis of case studies on early modern naturalists' networks, on networks of evolutionary biologists, of the marine station in Naples as a centre of communities of experimentalists, and the foundation of the European Molecular Biology Organisation lead to the following insights: First, from the early modern period on, we can identify transnational scientific communities of a self-described or implicit European character, though the sustainability and impact of these communities' Europeanness is uneven. Second, some of these communities contribute to a European identity and cause spillover effects of varying quality. Third, some developments exemplify the Haasian neofunctionalist logic of "spillovers" and help to explain how processes of scientific community-building can be understood as self-reinforcing processes. However, the shift of scientists' loyalties to a European level seems incomplete at best, since frequently scientists still need incentives in order to be brought to cooperate with partners from other European countries. Fourth, we find that in general, science turns to politics for funds, and politics to science for "innovation" and prestige. Overall, effects of science for promoting European identity exist but are quite limited.

A perspective from the sociology of science is presented by Julie Patarin-Jossec in her chapter on "Science and the manufacture of a political order: A sociological study of the European manned space programme". From a Marxist point of view, space flight may be seen as one of the most advanced manifestations of capitalism, transcending its realm beyond Earth towards space. Patarin-Jossec identifies three types of international coordination of activities, each of which is relevant for analyzing the European dimension of space projects: Distinguishing logistical, instrumental and structural cooperation leads to an improved understanding of how international collaboration in the management of European experiments aboard the International Space Station functions. Furthermore, Patarin-Jossec analyses the material bases for political power concerning space flight programmes. The focus lies on what she labels "bureaucratic caesarism", where a bureaucratic organization (here, the European Space Agency, ESA) exercises hegemonic political power in line with a paradigm of centralization of decision making. From a Marxist's point of view, a Europe of science, technology and research is the basis for the "fantasy" (Patarin-Jossec) of European integration. The work and status of ESA can be understood by taking into account the following four features: First, a constitution of scientific hegemony (monopolisation of scientific results), second, the logic of delegation, third, the proceduralisation of protocols (promoting credibility of scientific production), and fourth, a projection of territories into outer space as a way of dealing with a crisis of capitalism.

From a political scientist's point of view, Patricia Bauer analyses the formation of European Research Policy (ERP) from the perspective of neo-functionalism. In her chapter "European Research Programmes—Spaces of Knowledge or Economic Tool?", she points out the following crucial points of interdependency of integration policy and ERP: First, from its beginnings ERP has been promoted as a tool for integration. Second, to understand spill-over processes, it is heuristically fruitful to distinguish between bottom-up and top-down processes and correlated research agendas—mainly driven by the European Commission and by European Atomic

Energy Community (EURATOM) in the post-World War II era. Furthermore, Bauer reconstructs relevant stages of ERP, focusing on administrative procedures of organizing the policy as well as on how the top-down, centralized organization of ERP strengthens the role of the European Commission as a kind of pacemaker of research policy and of innovation. The programme of the European Research Area (ERA), set up in 2000, can be regarded as a milestone in opening "new pathways of networking, collaboration, pooling of financial means and industrial investment". Since this time, the concept of blurring borders of policies has been implemented consequently by the Innovation Union and the 8th Framework Programme, Horizon 2020. In sum, Bauer argues that, first, central views of research as a tool of integration have changed only marginally throughout the history of the EU. Second, neo-functional and market-oriented ideas in ERP promote the view that research policy enables and fosters integration processes. Third, a top-down mode of administration and agenda-setting in ERP is the maxim of defining what can be understood as relevant research. Fourth, basic research has to face severe acceptance problems, especially under the premise that basic research is not guided by maximizing economic payoffs.

In the following chapter, Stephen Rozée elaborates on the role of Europol and the EU Rule of Law mission in Kosovo (EULEX) in his case study. The chapter "Expert communities and security: Neofunctional integration in European policing practices" addresses two questions: First, how can we examine the formation of European expert communities in policing?, and second, to which extent can a neo-functionalist account explain the development of European integration in policing? According to Rozée, a wealth of findings suggests that—notwithstanding certain limitations—neo-functionalism essentially is a sound theory. First, the formation of EULEX and Europol can be grasped as a process of focusing the expertise of practitioners from EU states and achieve shared goals—i. e., promoting a European institution that fights crime. The development of EULEX and Europol can be understood by application of the neo-functionalist theory as products of functional spillover: In the case of Europol, as one effect of Schengen, the removal of national borders forced national police agencies to enhance national and transnational police cooperation. Analogously, a neofunctionalist explanation of the development of EULEX has to consider security aspects outside the EU. Rozée also detects some limitations of neofunctionalism, such as reluctance of EU member state to contribute to Europol or EULEX. Moreover, practitioners show a lack of commitment to EU policing mechanisms. Nevertheless, this does not imply that European integration in areas of counter-terrorism or fight against crime is not essential for security.

Sebastian Nähr contributes to this collection with his chapter "An epistemic-con-sequentialist social epistemology as an epistemological perspective concerning the investigation of a common, European knowledge community". The perspective of social epistemology (in the following SE) offers some insights to a deeper understanding of what a European space of knowledge might be. Sebastian Nähr identifies some weaknesses of Kusch's communitarian SE, i. e., Kusch cannot adequately protect his version of social epistemology against undesirable relativistic consequences. According to Nähr, Alvin Goldman's veritistic SE has to be confronted with three critical arguments. One of these counter-arguments rests on the assumption that Goldman's approach lacks descriptive clarity and depth. Subsequently, Nähr defends a version of an epistemic-consequentialist SE. In the framework of this theory, Europe is, after all, more than a historical, sociological, political etc. construction, Europe is an epistemic community. This thesis is justified by identification of common social-epistemic practices within an epistemic-consequentialist analysis. In detail, this means to fix the epistemic values that are regarded as prevalent and justified in a specified epistemic community. For example, if we consider the EU as an epistemic community, then different epistemic systems encompassing different epistemic practises are relevant, such as European law, the European political system and the European education system(s). Certainly, in each of these systems questions about what exactly knowledge qualifies as knowledge can be fruitfully analysed in an interdisciplinary manner. Relevant disciplines like sociology, cultural studies, social epistemology, and history of science offer an account that helps to describe and evaluate social-epistemic practises of different epistemic communities.

1.8   Acknowledgments

This collection is one result of the EU-funded research project "Europe's spaces of knowledge: Ideas, Discourses, and Integration of Elites in Science", carried out from 2014 to 2016 at the University of Siegen by the two editors, Bertold Schweitzer (Dundee) and Thomas Sukopp (Siegen).

Of course, a volume like the present one is not the product of two editors only. We are indebted to the contributors to the project (and this volume) who not only made the workshop in December 2015 and the conference in June 2016, both held at University of Siegen, a success, because of the prolific atmosphere and the lively discussions across disciplinary boundaries. Furthermore, we have to thank Frank Schindler and the team of the Springer publishing house, who patiently took care of the details of this publication. Our research assistants Falk Dreisbach and Andreas

Bender tirelessly provided valuable and competent support, both in preparing and managing the conferences as well as in assisting us in our editorial tasks.

We heartily thank the University of Siegen for their repeated hospitality, providing the venues for our workshop and conference. Most of all, we have to thank the European Commission for generously funding this project through the Erasmus+ Jean Monnet programme.

References

Anderson, Benedict. 2006. *Imagined communities: Reflections on the origin and spread of nationalism*. rev. ed. London: Verso.

Cahan, David. 2003a. Institutions and communities. In *From natural philosophy to the sciences: Writing the history of nineteenth-century science*, ed. David Cahan, 291–328. Chicago: Univ. of Chicago Press.

Cahan, David, ed. 2003b. *From natural philosophy to the sciences: Writing the history of nineteenth-century science*. Chicago: Univ. of Chicago Press.

Commission of the European Communities. 2007. *European space policy*. Communication from the Commission to the Council and the European Parliament COM(2007) 212. Brussels.

Dunlop, Claire. 2000. Epistemic communities: A reply to Toke. *Politics* 20: 137–144. doi:10.1111/1467-9256.00123.

Fleck, Ludwik. 1981. *Genesis and development of a scientific fact*. Edited by Thaddeus J. Trenn and Robert K. Merton. Chicago: Univ. of Chicago Press.

Haas, Ernst B. 1958. *The uniting of Europe: Political, social and economic forces 1950–1957*. London: Stevens.

Haas, Ernst B., Mary P. Williams, and Don Babai. 1977. *Scientists and world order: The uses of technical knowledge in international organizations*. Berkeley, CA: Univ. of California Press.

Haas, Peter M. 1992. Introduction: Epistemic communities and international policy coordination. *International Organization* 46: 1–35. doi:10.1017/s0020818300001442.

Holzner, Burkart. 1968. *Reality construction in society*. Cambridge, MA: Schenkman.

Kuhn, Thomas S. 1962. *The structure of scientific revolutions*. Chicago: Univ. of Chicago Press.

Popper, Karl R. 2002. *The logic of scientific discovery*. London: Routledge.

Toke, Dave. 1999. Epistemic communities and environmental groups. *Politics* 19: 97–102. doi:10.1111/1467-9256.00091.

Discoveries of Oxygen and the "Chemical Revolution" in the Context of European Scientific Networks

2

Thomas Sukopp

Abstract

The so-called Chemical Revolution in the late 18th century often related chemistry to Lavoisier's discovery of oxygen. It is argued that modern chemistry, and its rise as a scientific discipline, which began maybe from the 17th century but reached a kind of maturity with Lavoisier, or later, with Dalton, needs a more thorough analysis of its scientific and epistemic networks. Though a lot of detailed work devoted to the discoveries of "oxygen" has done, a more thorough understanding of the Chemical Revolution needs a deeper understanding of the following crucial issues: I) The interrelatedness of seemingly unconnected disciplines such as historiography of chemistry, philosophy of chemistry and sociology of science; II) The merits of different concepts of phlogiston theories as well as the rise of Lavoisier's chemical theories were transmitted and transformed in scientific networks. Analyses of how the inner circle of Lavoisierians—as a scientific community—exchanged ideas as well as adherents of phlogiston theories, such as Scheele and Priestley, lead us to the conclusion: Rejecting phlogiston theory as a whole is as irrational and fruitless as asserting that Lavoisier's oxygen theory is—in plain and simple terms—correct. III) We agree that in a very specific notion we are justified to label the changing chemical paradigms from 1770s to 1790s a chemical revolution. We will argue that Lavoisier, as the head of a scientific research group, was part of a chemical revolution for several reasons: a) revolution of vocational training for chemists, b) growing and intensified importance of manifold communication, c) chemical theory with unifying and explanatory power as a "package deal", and d) a special version of rationalism that was very much influenced by social factors and shaped a style of chemical reasoning.

The so-called Chemical Revolution in the late 18th century often related chemistry to Lavoisier's discovery of oxygen. It would be inappropriate to attribute the merits exclusively to Lavoisier, of course. But that is not the main point of this contribution. I will argue that modern chemistry, and its rise as a scientific discipline, which began maybe from the 17th century but reached a kind of maturity with Lavoisier, or later, with Dalton, needs a more thorough analysis of its scientific and epistemic networks. I assume, without further arguments, that compared with physics and biology, chemistry is a "late science". Nevertheless, "oxygen" is a case study that exhibits why chemistry has evolved as a scientific discipline since the late 18th century. This contribution has to be framed in a general concept of what chemistry is.

First, chemistry has always been a laboratory science (Gooday 2008, 788; Müürsepp 2016, 215; Meinel 2000, 287–302) and a practical science (Müürsepp 2016, 213–223). We should always keep in mind the word "labour" in connection with "laboratory" to understand that chemistry has evolved as an art that is related much more to craftsmanship than to science. The thinking style of chemists—to revert to Ian Hacking—is deeply influenced by the socio-cultural setting of laboratories.

The problems and objects related with the study of chemistry have been provided by and limited by the operations that could be performed on materials in a chemical laboratory [...]. As theoretical structures changed and new objectives supplemented or displaced the older ones, the stable setting of a chemical laboratory identified and distinguished chemists from other natural philosophers who dealt with natural phenomena in more theoretical and abstract way. (Holmes [1993], quoted in Bensaude-Vincent 2014, 332)

The practice of chemistry still needs physical activity as much as mental exercise.

Second, the term "Chemie" (chemistry) was first coined by Johann Joachim Lange in the middle of the 18th century. The Greek notion of "χημεία chimeía" means "the art of metal foundry". This concept points at transformation or transubstantiation, which refers to the transformation of a less noble metal into gold. Chemistry was an art in line with the meaning of the Latin word "ars" and also the Greek concept of "techne". It was not a natural science for a long time because natural laws and the abstract universal theoretical understanding of the inner secrets of nature lie in the area of physics.

Third, different academic cultures in different scientific communities located in different countries, and within a specific historical setting, shape chemical theories. For instance, vocational and educational practices, and the corresponding organizational structures in France and England from the late 17th century to the climax of the Chemical Revolution in the 1780s, have been found to be crucial for the formation of chemical theories.

The line of argumentation is as follows: In section 2.1, I will outline why and how the history of science is important for the philosophy of science with respect to systematic approaches. In section 2.2, I will reconstruct what Scheele, Priestley, and Lavoisier discovered. It seems to be clear that the discoveries of "Oxygen" tend to be ex-post explanations from the point of view of modern chemistry. To understand what these three figures thought they had discovered, and how they justified the related theories (oxygen or phlogiston), their scientific networks must be analysed (section 2.3). I will focus on the rather neglected aspects of their fabrication of knowledge. Section 2.4 will ask why we should or should not regard the long path of the finally-accepted oxygen theory as a chemical revolution. Again, many significant publications do not consider scientific networks as highly relevant.

The main thesis of this paper is that all prevailing interpretations of the philosophy of science fail in some respect because the so-called Chemical Revolution is much more complex than most philosophers of science think. The moderate optimistic conclusion is that we can understand knowledge formation processes much better if we concentrate on the analyses of (informal) meetings of scientists, correspondence, lectures and textbooks (see also Bensaude-Vincent 1990).

2.1   Why is the History of Science Relevant for the Philosophy of Science?

With respect to the so-called Chemical Revolution, it is obvious that the philosophy of science can actually learn some lessons from the historiography of science. Recent attempts of reconstructing the diverse shifts from phlogiston theories to oxygen theories show how standard philosophical explanations of theory shifts, i. e. the dynamics of competing theories, fail in different ways. We will examine these failures in section 2.4.

This section aims at pointing out how the philosophy of science can really learn from the history of science. This means that if the history and the philosophy of science are regarded as disciplines with equal rights from an interdisciplinary point of view, we can achieve better mutual understanding for both philosophical conceptualization and historical reconstructions. To put it in a nutshell, the history of science matters at least in the following respects for the philosophy of science:[1]

---

1   For the following two paragraphs, see Sukopp 2015, 242–43.

Philip Kitcher's[2] (2011, 505–524) statement that "(e)pistemology without history is blind" is of some value for our deliberations. Analogously, at least for heuristic purposes, the following statement will be fruitful to understand the changing views in the philosophy of chemistry: "The philosophy of science without history is blind." Briefly, Kitcher stated a threefold blindness relating to the ahistorical or anti-historical concepts in epistemology: First, the historical roots of philosophical problems are neglected or misconceived. Second, the fact that knowledge-bearers (e. g. scientists) belong to society—at least to a scientific community—and that knowledge is, at least in part, socially construed or distributed by social concepts, is declined. Third, to understand the paradigmatic changes in the establishment and the growth of knowledge, the status quo of knowledge in a given time has to be kept in mind but it is not done.

The general line of argumentation for a closer systematic linkage of these two disciplines with respect to the philosophy of chemistry follows recent insights into the history and philosophy of science-studies (HPS) (Stadler 2012; Ash 2012; Giere 1973). But it is also motivated by my conviction of classical philosophical analysis. As Richardson pointed out, HPS does well in mixing the history of science with the philosophy of science. Of course, the history of science is much more descriptive, while the philosophy is partly normative. Since Richardson analysed Carnap's "scientific philosophy", and since many philosophers of science are more or less adherents of a "scientific philosophy", e. g. naturalism, a historical reconstruction of naturalistic accounts, such as the theories of unity of science or reductionism in science (Gavroglu 1997) can foster our understanding of seemingly ahistorical views in the philosophy of science as shaped by historical settings.[3] An example that illustrates this argument is a standard interpretation of laws in science as eternal laws that fit into a deductive-nomological model. The laws that were regarded as the ultimate laws are those that can be found in theoretical physics. But chemical laws cannot be subsumed under this paradigm for several reasons (Bensaude-Vincent 2014, 60ff.; van Brakel 2014, 18). One reason is that the processes and structures of (theoretical) physics and chemistry differ from each other (van Brakel 1997, 263ff.). If chemists are interested in the transformation of substances, and if these transformations are of concrete matter, we might argue like Meyerson (Bensau-

---

2   I am indebted to Fabian Burt for his insightful summary of Kitcher's paper.
3   See van Brakel (1997, 254): "Current opinion, in particular that of a more general or philosophical nature, is clearly the product of the history of ideas, not determined by scientific observation. For example, views as to whether or not a clear distinction should be made between a chemical atom, molecule or substance on the one hand, and a physical atom, molecule or 'bit of matter' on the other [...] do not solely depend on 'timeless' data reported in a 'timeless' descriptive ideal language."

de-Vincent 2014, 64) that it is impossible and not really interesting to reduce the secondary qualities to primary ones. I will sketch three more arguments about why the philosophy and the history of science are mutually interdependent or, at least, they benefit from each other.

First, in favour of the realization of the systematic relevance of the history of science (and related disciplines) is the fact that the concepts of change (e. g. Benfey 2000)[4] are relevant for the philosophy of science. At least, they have been the subject of the history as well as the philosophy of science. But the various historical settings and (r)evolutions are still often neglected for several reasons: A) From the point of the philosophy of science, which claims that theories are timeless, the historical understanding of the concepts of change are irrelevant. B) From the perspective of realism, which simply states that phlogiston theory is false or does not explain anything (does not refer to the real world, etc.), socio-historical prerequisites can be ruled out.

Second, to refer to Richardson (2008, 99), "scientific ambitions of the logical empiricist philosophy of science can give historians a way of thinking about philosophy of science as something other than a dialectical opponent in the enterprise of understanding science". Why is this relevant? It is simply because logical empiricism is not only the result of history or philosophy but echoes scientific ambitions such as naturalistic attitudes (see also Cahn 2002; Ash 2012; Stadler 2012, 231, footnote 63).

Third, with respect to chemistry, we would like to stress that it is not only—and primarily—interested in the discovery of natural laws. It is rather interested in the analysis and synthesis of (new) compounds. Chemistry creates new substances. Furthermore, a shift from an analytical to a synthetic approach to chemistry is in close relation to the history of chemistry:

> The influential paper by Earley [...], although revolving around the problem of education, is probably the closest to my perspective on the issue. The change that is advocated in Earley's paper is described as a shift from an *analytical* to a *synthetic* approach to chemistry. The synthesis would be historical in nature: The old story-line of introductory chemistry courses—"whatever exists can be understood through *analysis* into its component parts"—is no longer sufficient. A more appropriate sto-

---

4   Benfey (2000, 200–1): "The most powerful concept that discipline after discipline has had to incorporate and cope with is the concept of change, and change is the pivotal concept of chemistry. For ours is the field of *real* change, where parts of the material world fundamentally change their nature. Shiny metallic sodium and poisonous green gaseous chlorine react to form common salt, the stuff of life. And we study not only reagents and products but more and more the actual process of change."

ry-line would be—"everything came to be through *synthetic* processes"—that is, the Evolutionary Epic. (Lamża 2010, 108)

There are several more arguments in favour of implementing historical reconstructions in the philosophy of science. Within the scope of this paper, I will leave them aside and turn to the following question.

2.2   Discoveries of Oxygen: What Did Scheele, Priestley, and Lavoisier Discover?

A simple and wrong answer is, of course, "oxygen". Oxygen was not conceptualized immediately as a modern chemical element. The epistemic status of "oxygen" as a basic substance that cannot be decomposed into simpler chemical entities was the result of later theories and deliberations. And scientists in interdisciplinary European networks strove for the best explanation, namely, the best metaphysical, physical, and chemical theories.

2.2.1   Carl Wilhelm Scheele (1742–1786)

This German-Swedish pharmacist and chemist (see Greenberg 2007, 291ff.) carried out various combustion experiments. He discovered oxygen in 1771/72. In modern terms, the crucial experiment can be described as the reaction of manganese dioxide and sulphuric acid producing manganese sulphate and oxygen:[5]

$$MnO_2 + H_2SO_4 \rightarrow MnSO_4 + \tfrac{1}{2}\,O_2 + H_2O$$

He discovered oxygen (O_2) by heating manganese dioxide (magnesia nigra; in Swedish, brunsten) with concentrated sulphuric acid (contemporary nomenclature: oil of vitriol). In his experiment description, Scheele wrote (1780, 43–44):

> I mixed so much of concentrated oil of vitriol with manganese reduced to an impalpable powder, that it became a thick, stiff *magma* (or dregs). This mixture I distilled in a small retort by open fire. In lieu of a receiver I employed a bladder emptied of

---

5   Those who would like to become more familiar with the chemical reactions of inorganic chemistry (e. g. redox-reactions or thermolysis) might want to consult well-known textbooks like Holleman et al. 2007.

air; and, that the rising vapours might not attack the bladder, I had poured into it some milk of lime [...]. As soon as the bottom of the retort became red hot, a kind of air passed over, which gradually expanded the bladder. This air has all the qualities of a pure *pyreal air*.[6]

Adhering to phlogiston theory, he called it "fire air". Light, according to Scheele, consists of "flammable air" and "phlogiston", which is exhausted during combustion. In 1774, he reported his discovery to the Royal Society in Stockholm. His results, among them along with the thermal decomposition of quicksilver oxide into quicksilver and oxygen, were to be published in 1775. But it took another two years before Scheele published his findings in his famous *Chemische Abhandlung von der Luft und dem Feuer*[7] (Scheele 1777). He discovered a part of air that promoted combustion, which he called "fire air" or "pure fire air" (Ladyman 2011, 90).

It is widely accepted that Scheele was at least at that time a strict adherent of the phlogiston theory (Stewart 2012, 177ff.). What we call "air" consists of oxygen (Scheele: "fire air" or "pure fire air") and "mephitic air"[8] (in modern terminology, nitrogen). To give a first hint of what phlogiston is, the following passage is quite instructive (Scheele 1780, 13):

It appears from all these Experiments, that in each of them phlogiston, the simple inflammable principle, is present. It is well known, that Airs attracts the inflammable parts of bodies, and deprived them of it: not only this may be seen from the above Experiments; but it also appears that in the transition of what is inflammable principle into to the Air, a considerable part of the Air lost [...].[9]

---

6   Original: "Ich mischte so viel concentrirtes Vitriol Oel unter zart geriebenen Braunstein daß es wie ein dicker Brey wurde. Diese Mischung trieb ich aus einer kleinen Retorte in offenes Feuer. In der Stelle eines Recipienten gebrauchte ich eine Luft leere Blase und damit die etwa übersteigende Dünste die Blase nicht angriffen, so hatte ich etwas Kalkmilch in selbe gegoßen [...]. So bald der Boden der Retorte glühete, gieng eine Luft über, welche die Blase nach und nach aus dehnete. Diese Luft hatte alle Eigenschaften einer reinen Feuer Luft" (Scheele 1777, 35–36 (§ 32)).

7   In English: *Chemical observations and experiments on air and fire*.

8   The terms that are in quotation marks can be found quite often in Scheele 1777 (engl. 1780). Already in his preliminary report, p. 4, Scheele makes use of the term "Feuer-Luft" "fire air", "pyreal air"; see also p. 35); similarlly, on p. 4 he speaks of "verderbter luft" ("mephetic air"), resp. on p. 122 (§89) von "verdorbener Luft".

9   Original: "Man siehet aus diesen Erfahrungen, daß bey jedem Versuche, das Phlogiston, dieser einfache brennbare Grundstoff, zugegen ist. Man weiß, daß die Luft das Brennbare der Körper starck an sich ziehet und selbiges ihnen beraubet: dieses ist nicht allein aus angeführten Versuchen zu sehen, sondern es erhellet zugleich daß bey dem Uebergange des Brennbaren in die Luft ein mercklicher Theil Luft verlohren gehe" (Scheele 1777, S. 11 [§ 15]).

- Phlogiston is exhausted in combustion processes.
- Phlogiston is a simple, flammable, elementary chemical substance (for further notions of phlogiston, see Allchin (1997, 473–74)).

So, what did Scheele discover? He thought he had discovered a part of air that promotes combustion. He called it "fire air". His findings were subsequently integrated into the theoretical framework of the phlogiston theory[10] *(see also sections 2.3 and 2.4). He did not know anything about the independent discovery of oxygen by Priestley.*[11]

2.2.2   Joseph Priestley (1733–1804)

The British-American theologist, physicist, philosopher, and chemist, Priestley, started his career in England before he immigrated to America. Among other important contributions to chemistry, he synthesized various nitrogen oxides and soda water. He was promoted by Benjamin Franklin and was a strict adherent of phlogiston theory until he died. He discovered oxygen independently two years after Scheele but published his results in 1774. In his main chemical work *Experiments and Observations of different kinds of Air* (six volumes), and also in the early volumes of *Philosophical Transactions*[12], Priestley termed oxygen as "dephloginized air"[13] or "dephlogisticated air" (Ladyman 2011, Stewart 2012).

---

10 See e. g. Scheele 1777, Vorbericht, S. 2ff. or S. 24 (§ 27), S. 70–71 (§ 66)

11 Torbern Bergman (Scheele 1777, Vorbericht, S. 16): "Schließlich muß ich noch erwähnen, daß dieses meisterhafte Werk schon bey nahe zwey Jahre fertig gewesen, ob es gleich wegen mehreren Ursachen, die hier anzuführen überflüßig sind, erst jetzt herausgegeben wird. Hiedurch ist es aber geschehen, daß H. Priestley, ohne von des Hr Scheele's Arbeit zu wissen, noch vorher unterschiedliche neue Eigenschaften, die der Luft eigen sind, beschrieben hat. Man sieht sie aber hier so wohl auf eine andre Art, als in einem andern Zusammenhange vorgestellet." ("Prefatory Introduction" to Scheele 1780, xl: "Finally, it is necessary to observe, that the following interesting performance, has been ready for the press very near two years ago; though it is published only at present, for several reasons, which it would be superfluous to alledge here. By this delay it happened, that Dr. Priestley not knowing any thing of Mr. Scheele's work, has previous to its publication, described several new properties of Air. But they are here exhibited in a different manner, as well as in a quite different connection."

12 Priestley 1774, 1775, 1777, 1779 (Vol. 1–4); 1781, 1786 (Vol. 5–6), 1790 (Complete edition, 3 Vol.)

13 Priestley 1775 (Vol. 2, Section III: Of Dephlogisticated Air, and of the Constitution of the Atmosphere, 29–61. (see also sections IV and V. The titles of both sections contain

Priestley produced oxygen through thermal decomposition of quicksilver oxide, according to the following chemical equation:

$$2 \, HgO \overset{\Delta}{\rightarrow} 2 \, Hg + O_2$$

He was the first to publish reports on the positive effects of oxygen on respiration processes. He wrote that this pure air would be a kind of luxury item in future. He mentioned that only he and two mice had so far had the privilege of breathing that pure air.

2.2.3   Antoine Laurent Lavoisier (1743–1794)

The chemical function of oxygen in relation to the processes of combustion could not be explained sufficiently because two rival paradigms tried to explain combustion.[14] Lavoisier was an early proponent of the oxygen theory of combustion, whereas Priestley and Scheele believed in the superiority of the phlogiston theory. Lavoisier figured out that the products of combustion were heavier than the reactants. Though this had been a fact for a long time, the exact weighing of the reactants and products was rather new. Lavoisier hypothesized that oxygen was the reason for the increased mass of combustion products. He also decomposed iron oxide (Fe_2O_3) in 1774 and got oxygen, but did not think of oxygen as a chemical element. In the autumn of 1774, Priestley reported (Jansen 1994, 162) his experiments to Lavoisier in a letter. These helped Lavoisier's own experiments a lot. In 1775, he carried out crucial experiments and thought that oxygen caused the increase in the mass of calcinated metals. Lavoisier also pointed at the eminent role of oxygen in the context of vital processes.[15] At that time, oxygen was thought to be an essential sort of acid. In 1779, Lavoisier proposed the term *Oxygenium* (French: oxygène, i. e., generator of acids; Ströker (1982, 239–40)).

---

the term "Dephlogisticated Air"). Priestley carried out various experiments with this gas (see e. g. Willeford 1979, 111–117, here 113).

14   Carrier 2009, 12–42, here 12 ff. and "Chapter 2: Scholarly person and tax farmer and early career".

15   Oxygen became a mass product of chemical industry in early 20th century. Carl Linde had invented oxygen condensation in 1877.

2.3   Scientific Networks in the Context of the Chemical Revolution

Discoveries of oxygen in the context of oxygen theory or of phlogiston theory are an interesting example of diverse reconstructions of the so-called Chemical Revolution. These reconstructions are more or less interdisciplinary in nature, since at least four disciplines are relevant for the analysis of prevailing scientific theories and their epistemological, ontological, and methodological attitudes: The philosophy of science, social epistemology, the sociology of science, and the history of science.

I will focus on a key aspect of scientific networks. The complex processes of adopting and rejecting phlogiston theory can only be understood when we analyse the various communication and collaboration relations of the actors. While doing this analysis, we will also look into how the different styles of communication frame scientific communities in different countries. Though it is widely accepted that scientific communication and collaboration are important, it is also commonly accepted that the official gold standard of science is a treatise, an opus magnum, etc. I will partly reject this view for the following reasons.

2.3.1   Elements of Phlogiston Theories

The history and philosophy of phlogiston is still a rich field of controversies (Boantza and Gal 2011, 317–342; Chang 2010, 47–79; Woodcock 2005, 63–69; McEvoy 2010; Ladyman 2011; Stewart 2012). I will summarize some key aspects of phlogiston with respect to the different communication and collaboration practices[16]. To start with, there was no single, widely accepted ontology in chemistry in the late 18th century (Stewart 2012). Stewart referred to the works of Mi Gyung Kim. She pointed at three primary roles of phlogiston. The first is the identification of phlogiston with the sulphur principle (Stewart 2012, 176). Wilhelm Homburg (1652–1715), a leading

---

16   See e.g. Stewart (2012, 179ff.), who referred to Georgette Taylor's analysis of lecture notes written by Will Falconer (1732–1769) of textbooks: "In these lectures, Cullen proposed a novel, gas-based identity for phlogiston. He said that phlogiston was not a simple, indivisible element but rather a chemical compound composed of an acid and mephitic air. Cullen based this theory on the empirical observation that any object when inflamed releases a noxious/ mephitic form of air that does not support animal life and that the remaining ash or calx is acidic. Noting that phlogiston had never been isolated, Cullen concluded that phlogiston's constituent mephitic air and acid did not have a high enough affinity to combine with each other unaided, but rather joined only when a third substance such as a metal or earth was introduced."

scientist of the *Academie Royale des Sciences* in Paris who corresponded on that issue with Robert Boyle (1627–1691), identified the sulphur principle as "the active principle in all bodies" (cit. after Stewart 2012, 178). Phlogiston's second identity, according to Kim, was derived from this operational role in analysis and synthesis of metals. For Geoffroy, phlogiston was a "concrete oily substance separated out in chemical analysis" (Kim 2008, 34). The third theoretical identity of phlogiston was developed by Louis Lemery (1677–1743). Phlogiston, according to Kim, is the "first and most powerful solvent", which at times "failed to dissolve solid bodies and became imprisoned in them" as fixed fire (Kim 2008, 35). According to Kim,

> Lemery tied together the theoretical, ontological, and operational identities of phlogiston into a coherent and stable system that became the standard interpretation of phlogiston in French chemistry for the next fifty years. (Stewart 2012, 178)

Obviously, it matters which of these identities is regarded as the prevailing identity. It is striking that British chemists favoured the view that the operational identity is the most important feature of phlogiston and reinvented this identity through pneumatic chemistry.

From a modern perspective of scientific realism and according to the correspondence theory of truth, it is simply wrong that a substance "phlogiston" exists. There is no substance that exhausts via combustion. The thesis that phlogiston is a "principle of earth", i. e., a variety of the element earth, as put forward by Georg Ernst Stahl (1660–1734), could be labelled as quirky or abstruse. But phlogiston had an enormous explanatory potential (Stewart 2012; Carrier 2009; Labinger and Weininger 2005, 1950–51)[17]. Phlogiston helped integrate different phenomena in one theoretical framework. It had a kind of unifying power. All combustion processes, as well as all calcinations of metals (to calces of metals; in modern terminology, metal oxides), could be explained under the paradigm of phlogiston. "This was by no means obvious. It is important to note that until the mid-eighteenth century, gases arising from combustion, such as carbon dioxide, were simply seen as 'airs' and not collected" (Greenberg 2007, 331).

When we assume an explanatory entity, "phlogiston" (Frercks 2008, 383), what did it explain? It could not only be hypothesized that phlogiston is exhausting (as a flammable gas, i. e. the flammable component of a substance that undergoes the process of combustion). A) Phlogiston also gives metals their characteristic brilliance, plasticity, and electrical conductivity (Frercks 2008, 383). B) Phlogiston

---

17 It is quite remarkable that even chemists like Linus Pauling, who normally argue in a rather ahistorical manner, ascribe explanatory power to phlogiston theory.

theory explained the effects of phlogiston because of its movements in space. C) Phlogiston could be considered as a unifying theory with respect to explaining not only combustion and calcinations but also oxidation (phlogiston is exhausted) and reduction processes (phlogiston is chemically bound). D) Phlogiston theory helped explain qualitative descriptive features of chemical processes. It could be easily adopted for qualitative changes through chemical processes (Ladyman 2011, 98). E) Phlogiston also helped create a lot of new experiments (see Priestley, on how it contributed to modern chemistry, see Willeford (1979, 112–13)). F) Though phlogiston theory had to cope with difficulties[18], compared with oxygen-oxidation theories, phlogiston theory could be regarded as successful even in the 1780s. The most striking deficiency of phlogiston theory was that since 1730 (Rey), it was well known that combustion products were heavier than the reactants (e. g. metal plus oxygen). So, it could be regarded as pre-scientific.[19] But, how then could it be any more successful?

2.3.2 One Success of Phlogiston Theory and Communication Networks

Cavendish and Priestley thought that phlogiston theory was splendidly verified (Carrier 2009, Section 4). They identified phlogiston as equal to flammable air. The following experiment consolidated this thesis: in modern terms, hydrochloric acid (HCl) reacted with iron (Fe) to form iron chloride ($FeCl_2$) and hydrogen (H_2).

$$2 \, HCl + Fe \rightarrow FeCl_2 + H_2$$

Cavendish claimed that the characteristics of flammable air were independent of the nature of the reacting acid. He consequently hypothesized that the gas (modern: hydrogen) could be no chemical part (component) of the acid. Thus, it had to

---

18 Ströker (1982, 250) summarized that the contradictory attributes of phlogiston as a substance led to severe theoretical problems for phlogiston theory. Phlogiston was conceptualized as an imponderable and weighable substance. It could penetrate vessels. Sometimes it was regarded as a matter of light or as caloric (matter of heat). Moreover, the theoretical basis of phlogiston in terms "principle" and "affinity" were unclear (Ströker 1982, 240–41; Frercks 2008, 210). Nevertheless, Lavoisier did use elements of phlogiston theory, e. g. he termed oxygen as the "principle" of acids (Ströker 1982, 240).

19 The positivistic camp was satisfied with labelling "Phlogiston" as an illusion. Gale stated: "Science, according to the positivists, simply had no need to become involved with such illusionary entities" (Gale 2003, 615).

be part of the metal. This gas he named "pure phlogiston"[20]. What could Lavoisier do? In his concept, all metals were elements. Therefore, no flammable gas could be part (component) of the metal. He, also in his view consistently, concluded that the flammable gas is part of the acid. If the flammable air (modern: hydrogen) is part of the metal, then it should recombine to form acid (modern: hydrogen plus oxygen must react to acid). All respective experiments that Lavoisier carried out failed. He correctly knew about the role of oxygen in the oxidation processes but misunderstood the role of oxygen as an essential part of all acids (oxygenium=generator of acids). HCl (hydrochloric acid) is an acid but does not contain oxygen. The reaction of oxygen and hydrogen yields water, which is no acid.[21]

> Lavoisier and I have repeated recently before Mr. Blagden and several other persons, the experiment of Mr. Cavendish upon conversation into water of dephlogisticated and inflammable airs, by their combustion; with this difference, that we have burned them with assistance of electric spark, by bringing together to currents, the one of pure air, the other of inflammable air. We have obtained in this way more than 2½ drachms of pure water, or which, at least, had no character of acidity, and was insipid to the taste. (Letter from La Place to Deluc, 28 June 1783; cit. after Thorpe 1921, 42)

So far, it is not clear how and why the analysis of the social dimension matters. It should be clear that no one could, at a glance, reject phlogiston theory and adopt oxygen theory (Lavoisier lectured to the *Academie* in Paris rather late, in 1783, on the abandonment of phlogiston). Also, a clear and exclusively predominant confrontation, labelled "oxygen theory versus phlogiston theory", is inappropriate. We have to consider both as package deals of bundle theories. For the contemporaries from the 1760s to the 1790s, it was neither possible nor plausible to make a simple case against or in favour of one of these theories. The clear "either-or" is an oversimplification, as we will see when we focus on correspondence and communication networks. First, it was a topic of many controversies—even between the adherers of phlogiston theory, regarding what phlogiston was. (I have sketched three aspects of phlogiston above.)

---

20 "Cavendish would later explicitly deny his original claim that phlogiston was inflammable air." (Stewart 2012, 180)

21 Ströker (1982, 241) remarked that Lavoisier was wrong on this point but followed an inner logic of oxidation theory. This theory had to assume that oxygen is an essential part of all acids: "The oxidation theory demanded that in addition to analysing all known acids to verify the oxygen principle, the acids be synthesized by means of the oxygen principle. This procedure should help to gain insights about the chemical composition of acids."

It is an oversimplification to speak of *the* phlogiston theory (Ladyman 2011). As we have argued, the way to grasp the meaning and chemical, operational and ontological dispositions of phlogiston was highly controversial: A) Early-stage phlogiston theories thought that phlogiston was a theoretical entity that could be identified in concrete chemical compounds. B) Cavendish termed hydrogen (a product from the chemical reaction of hydrochloric acid with iron or other metals; see figure 3; Ströker 1982, 191) as phlogiston. C) In this context, phlogiston was understood as the principle of combustibility (Ströker 1982, 192; Chang 2010, 70). D) Phlogiston was also termed "fire substance" (Ströker 1982, 265).

I will now present three examples of scientific communication strategies with respect to chemists who were proponents of the phlogiston theory.

First, we can learn from the different notions of phlogiston theory that, for example, Kirwan and Cavendish, as adherers of a certain version of this theory, and Lavoisier, who was, of course, one of its opponents, agreed on the empirical definition of inflammable air:

> In pneumatic chemistry, phlogiston made the transition from metaphysical principle to empirically defined, chemically irreducible element. By 1783, inflammable air was a chemical substance with a stable, broadly accepted operational identity. Kirwan, Cavendish, and Lavoisier would all agree that it was the product of a reaction between acids and metals, identifiable by its low specific gravity and its explosiveness. However, those three men held three different ontological identities for inflammable air. Kirwan thought it was a chemical simple, the aerial form of phlogiston. Cavendish thought it was a compound of water and phlogiston, and Lavoisier thought it was a compound of hydrogen and caloric. (Stewart 2012, 181)

They did so because of their shared views on valid arguments (Stewart 2012, 180–81) and because, for example, Cavendish could read a translation of Scheele's works (English translation 1780, with notes by Kirwan) and presented papers to the Royal Society in London (in 1766 and 1784). The communication here was between equal-ranking men of science of nearly the same status.

Second, the correspondence of Scheele, in combination with his specific scientific socialization, sheds light on how his works were valued and how he shared his knowledge:

Our argument rests on the assumption that Scheele socialized in a special manner as a scientist and this socialization had an impact on his argumentation style[22].

---

22 See Gross, Harmon, and Reidy (2002, 101–2) for an analysis of arguments in the context of phlogiston theory. In short, chemical explanations are different from explanation in other sciences. See Gross, Harmon, and Reidy (2002, 75ff) for an analysis of Lavoisier testi(fying) to the bonds of politeness among French scientific gentlemen as firm as those

Fors (2003, 167–190) had analysed in detail the correspondence of Scheele in a network of chemists, metallurgists and pharmacists, especially in the context of the *magnesia nigra* (modern: magnesium dioxide) experiments. For our purposes, a brief argumentation can be derived from the insights that Fors gave us: A) The traditional view of Scheele, that he was indeed a genial experimenter but had rather poor theoretical knowledge, has to be revised from the perspective of his rich correspondence (Fors 2003, 18). B) The correspondence shows the following characteristics: Bergman, the most respected chemist in Sweden, had this success because of an adoption of innovations from Britain and France and because of the "work and expertise of Carl Wilhelm Scheele and the chemists at the Board of Mines, who, among other things, produced new, previously unheard of substances, that could be named and ordered into systems […]" (Fors 2003, 105). Bergman was well aware of the capacity of Scheele as a scientist. But, Bergman held a much more prestigious position than Scheele, who started his career as an apprentice in a pharmacy. Gahn functioned as a kind of mediator between Bergman (his teacher and patron) and Scheele. Bergman and Scheele could not communicate as equal-ranking persons. C) The correspondence in the context of the *magnesia nigra* experiments shows how Scheele had to lower his status to communicate with the eminent and prominent university-chemist Bergman. But, a consequence of Scheele's lower social status was that he exclusively wrote about chemical experiments and chemical theories (Fors 2003, 183). That helped him very much to become a socially and scientifically accepted chemist who showed his chemical abilities through his correspondence and "placed the receiving correspondent in a position of scientific dept" (ibid.). D) The correspondence about Scheele's first important chemical paper "On Brown-stone or Magnesia nigra and its properties" ("Om Brun-sten eller Magnesia, och dess Egenskaper" (1774)) helped him boost his social and chemical reputation. The paper contained, among other findings, discoveries of christened chlorine, manganese, and barium. He was regarded as an excellent experimentalist who "presented his claims in a clear and polite manner" (Fors 2003, 188).

Third, there is a difference in communication when we compare English and French scientists from the 1760s to the 1790s. However, the simple opposition demonstrated in pitching the egalitarian, individualistic thinker Priestley against the communitarian Lavoisier seems inadequate (for this view, see Nordmann 2002). It is true that the famous scientific societies (Crosland 2005, 25) mirrored a main difference in the organizational structure of both countries. While the Paris

---

between English scientific gentlemen. Lavoisier offered for example the virtual witness of his laboratory notebook as a king of moral instrument: He is able and will reproduce his experimental results.

Academy of Science was a state-sponsored and state-regulated organization with strict hierarchies and few members (around 50 in the late 18th century), the more egalitarian Royal Society in London had 500 members at that time, most of them amateurs. Competitiveness was a crucial element of the Royal Society in Paris. Though we do not agree with Crosland (2003, 334), who stated that the French chemists, "unlike some of his distinguished British contemporaries, such as Henry Cavendish and Joseph Priestley [...], did not work alone", the crucial point is that Lavoisier and his colleagues formed a research group with a research programme. They were working as a team, "possessing enlightenment, knowledge and means that would have been impossible to find in isolated individuals" (Lavoisier, quoted in Crosland 2003, n. 7). Crosland identified five criteria for research groups that were invented independent of the Lavoisier case (Crosland 2003, 336). Two of these criteria are very closely connected to communication practices: "(3) Preferably, a common workplace/laboratory with suitable materials and apparatus and many opportunities for informal discussion. (4) Access to a common organ for publication" (ibid.).

In fact, from 1775, Lavoisier worked in the Paris Arsenal, one of the best-equipped laboratories in France. Madame Lavoisier described the atmosphere of this place as follows:

> Some enlightened friends, some young men, proud to be permitted the honour of collaborating in his experiments, gathered in his laboratory in the morning. It was there that they had lunch, that they had discussions, that they worked, that they carried out experiments which gave birth to that fine theory which has bestowed immortality on its author [...] The very best craftsmen were admitted to these communal occasions to [plan and] construct the machines that Lavoisier commissioned [...] (quoted in Crosland 2003, 338)

The organ for publication was the *Annales de chemie* and many more journals. The impact, in terms of readers and dissemination, was enormous, as Bret (2016, 125ff.) has persuasively argued.

2.4 Reconstructions of the Chemical Revolution: From the Knowledge of Chemists to Philosophical Interpretations

With respect to the so-called Chemical Revolution under consideration, of which Lavoisier is the main figure, I will outline three theses that require further explication (section 2.4.1). In section 2.4.2, I will sketch a view that claims that we should term the developments from the 1760s to the 1790s as a chemical revolution. Still, my answer will focus mainly on communication strategies and scientific networks.

2.4.1 Three Theses in the Context of the "Chemical Revolution"

The first thesis is also the most important one:

Thesis 1: *Taking into account the fact that there are hundreds of publications on the so-called Chemical Revolution, we do not have a theory that tries to integrate results from various disciplines and have no clear understanding of what Lavoisier and phlogistonists claimed to have known with respect to knowledge formation in scientific networks.*

The following table will list the views that respect the so-called "Chemical Revolution", which is closely related to chemistry from the 1760s to the 1790s, as a scientific revolution. Only four researchers share a clear, positive answer to this question. The majority holds a kind of intermediate position, which does not mean a polite abstinence of judgement. Thus, there are, according to the majority, good reasons why Lavoisier was—more or less—the initiator and completer of a scientific revolution.

Is it a scientific revolution?

Yes, chemical	Yes, physical	Intermediate Position	No	Question rejected
Beretta (1993)	Melhado (1985, 195–211)	McEvoy (2000, 47–73): Sceptical with respect to a philosophical interpretation of "revolution"	Gough (1988, 15–33)	Bensaude-Vincent (1983, 53–178)
Crosland (1995, 101–108; 1980, 389–416)		Holmes (1988, 82–92; 1997, 1–9): Partial revolution		
Donovan (1988a, 5–12; 1988b, 214–231)		Perrin (1988a, 53–81)		
Chang (Kusch 2015, 73ff.)		Siegfried (1988, 34–50)		
		Stolz (1991, 46–50)		
		Kim (2005, 167–191)		

A synthesis[23] of the following four crucial questions demands future research: A) Are the developments in chemistry from 1770 to 1790 best explained as a scientific revolution? The unanimous position, e. g. of Kuhn, is not acceptable from our point of view. B) What is a chemical revolution? (We cannot focus on this question in this paper.) C) If we are right to speak of a Chemical Revolution, what are the relevant periods? D) What was the role of Lavoisier within the Chemical Revolution?

Ad A) We cannot focus in detail on the above-mentioned positions (see table 1). Without going too much into detail, we take a closer look at those four researchers, who agree that we should consider the developments as a scientific revolution:

---

23 Different researchers emphasized that a synthesis is a desideratum though there are numerous brilliant case studies (Frercks 2008, 339; McEvoy 2007, 605).

Lavoisier as a scientific revolutionary

Researcher	Scientific revolution because …	Appraisal
Evan Melhado	It was a revolution in physics. Lavoisier had a genuine interest in physics (see his experiments on "Airs" with respect to which we call aggregation states and his interest in caloric [substance of heat] as an element).	This view neglects what we can label as "genuinely chemical" (for the multifaceted research programme of Lavoisier, see, e. g., Frercks 2008, 341).
Arthur Donovan	a) Chemistry was no scientific discipline before the era of Lavoisier; b) Lavoisier's methodology was conceptualized based on experimental physics but adopted for the purposes of chemistry; c) this view, that we should phrase the developments of a Chemical Revolution, corresponds to Lavoisier's own view.	There are good reasons to state that the meaning of quantification in chemistry cannot be fully valued until Dalton (1766–1844) (see Siegfried 1988, 34–50). Apart from that point, Donovan put forward a lot of good arguments.
Marco Beretta	We should, according to strict historical standards, and with respect to Lavoisier's views and those of his contemporaries, clearly identify the chemical developments from 1770 to 1790 as a chemical revolution.	The crucial question is whether Lavoisier's understanding of "chemical revolution" is consonant with the modern concepts of "chemical revolution". Even if we find a consonance, this could be rather arbitrary, i. e. Lavoisier could not have acted fully intentionally in favour of a Chemical Revolution. In fact, he expressed ambiguous attitude towards his own "revolutionary" ambitions, as we will see later in this chapter.
Maurice Crosland	Lavoisier's activities, guided by the policy of science, clearly favour this answer. Lavoisier was engaged in objectification strategies of chemical language/nomenclature as well as in the objectification of chemical experiments in terms of a strict quantitative method.	This answer is motivated by the considerations of the sociology of science. The answer could—from our point of view—be integrated into a philosophy of science perspective (see Thagard 1994, 642–44).[24]

---

24 Thagard (1994, 644) also encouraged researchers to improve interdisciplinary approaches, as we have done in chapter 1 of this contribution: "A key conclusion to draw from the interdependence of cognitive and social scientific change is the appraisal of cognitive and social strategies must also be linked. Cognitive appraisal should take into account the fact that much scientific knowledge is collaborative, so that we should evaluate particular cognitive strategies in part on the basis of how well they promote collabora-

Ad C) and D): Even if it sounds a bit presumptuous, I will summarize what Lavoisier's inner circle claimed to have known. One of the essential collective epistemic convictions was that oxygen is elementary and that its "nature" becomes apparent in the processes of calcination and combustion. Furthermore, though not every adherer of Lavoisier's theory shared the view that phlogiston does not exist, most followed him in rejecting the existence of the substance "phlogiston".

The need for a much less essentialist nomenclature was also a strong impetus for Lavoisier to promote his views.

The exact chemical role of oxygen was rather controversial. While the adherents of the phlogiston theory, such as Macquer and Berthollet, thought that oxygen was a component of air, Lavoisier's followers disagreed on whether there were different chemical types of air. (We know that there is only one mixture of gases called air. But in the 18th century, you could not be so certain.)

It is a myth that Lavoisier was revolutionary in stressing the relevance of stuff/mass-balances. But the will to carry out more exact experiments than his opponents, combined with a non-revolutionary attitude, fostered Lavoisier's success. By the way, if these processes could be labelled as "Chemical Revolution", it took at least 20 years for some of Lavoisier's theories to be accepted. Even his critics were convinced that Lavoisier could reproduce experimental findings more correctly than most phlogistonists and that his theory was simpler than the phlogiston theory.

Finally, Lavoisier did not think that his theories were completely new or even revolutionary.

Thesis 2: The history of chemistry is still dominated by three main paradigms that cannot adequately understand certain aspects of knowledge formation in scientific communities.

Following McEvoy (2000, 47–73) and Lavoisier ([1789] 2008, 336–37) in this point, we find these paradigms: A) Positivist-whiggish (the lonely genius is paradigmatic for successful research. He, the genius, is always trying to discover nature's secrets. These slogans, of course, paint a somehow distorted picture. But the neglect of social elements in epistemology and in the philosophy of science is still very influential). Proponents of this view are, for example, George Sarton, Herbert Butterfield, Charles C. Gillespie (and Maurice Crosland with respect to the history of chemistry). What is wrong with this view? As we have seen, Lavoisier emphasized that he is part of a group of researchers. The focus on a single person is not very promising. Without

---

tion. Conversely, social appraisal should take into account the cognitive capacities and limitations of the individuals whose interaction produces knowledge."

doubt, Lavoisier played a prominent role. But to claim that there is one single shift in one single paradigm is—at least with respect to the complex change of paradigms, as we have already argued—a failure. And obviously, the change was not so rigorous and rapid, as a positivist-whiggish concept proposes. B) The post-positivist view (e. g., Lakatos, Kuhn, Popper, Donovan), which claims that the development of theories is more important than seemingly monolithic theories. Thinking about conditions for evaluating theories, which would take into account the historical setting of theories, is also important for theorists like Lakatos and Kuhn. What is wrong with this is stated as the following: apart from many important findings of, for example, Kuhn, the thesis of incommensurability and the rapid change of paradigms cannot be found or are at least questionable with respect to the reconstruction of Lavoisier's chemical theories (Ströker 1982; Chang 2010, 47)[25]. C) Some kind of postmodern historiography of science: the impact of social factors certainly marks important progress to explain the changes of scientific findings and scientific progress. But the prize of postmodern theories is, in my view, too high: the neglect of universal standards and the abandonment of truth and other valuable criteria do not fit our view of science (David Bloor, Stephen Shapin, and Bruno Latour seem to give up those criteria). To abandon certain universal standards is a kind of capitulation of reason.[26] A possible consequence of postmodern historiography and postmodern philosophy of science is to reject the search of methodological standards and universality in terms of similarities and common grounds for paradigm shifts, scientific methods, and scientific progress.[27]

---

25 For a view that argues in favour of Kuhn, see Hoyningen-Huene (2008, 101), who summarized: "The result will be that Kuhn's general description of scientific revolutions fits the chemical revolution extraordinarily well." From our point of view, Hoyningen-Huene argued consistently to a certain degree but conceded that the Kuhnian incommensurability thesis (here matured phlogiston theory versus oxygen theory) cannot be applied to the chemical revolution: "We can say with the advantage of hindsight that the concept of a chemical principle no longer had any substantial role to play and was about to be banished from chemistry. This did not happen in Lavoisier's times when oxygen was thought of as the principle of acidity, but it did happen in the early nineteenth century […]" (2008, 112).

26 "While thus opposed to the essentialist and historist vision of unity, linearity, and homogeneity of a single, absolute time, Althusser and Foucault also rejected the pluralistic orientation of the Annales authors, who affirmed 'the existence of different temporal strata and rhythms—the political, the economic, the geographical—without attempting to establish any systematic links between them' (Dews 1994, 112)" (McEvoy 2000, 61).

27 Even if a naive concept of "progress" is rightly discredited, we could and should ask if rival theories can be compared with each other and, thus, a theory A is superior to theory B in terms of explanatory power, theoretical consistency, etc. Of course, this does not

Now, what I suggest is something like a pluralistic account of social epistemology, a social philosophy of science, and a social historiography of science in a truly interdisciplinary manner. What we need is the following: The existing accounts of the Chemical Revolution ...

> are not *historical* to the extent that they subsume history under the disciplinary interests and categories of science, philosophy, or sociology. Instead of grasping the Chemical Revolution as a product of history, a specific mode of temporality, they view it as a scientific discovery, a moment of rationality, or a matrix of social interests, which happen to have occurred in the past [...]. In order to evaluate these disciplinary intrusions, and to answer Cohen's question "*or What?*" with an unequivocal "*History*", we need to develop a clear sense of the irreducibility and specificity of history and the methods used to study it. Instead of approaching history scientifically, philosophically, or sociologically, we need to treat science, philosophy, and sociology historically. Instead of viewing chemistry as a well-defined science, "which has a history one can choose to study or ignore", we should, as Bernadette Bensaude-Vincent and Isabelle Stengers suggest, "envisage this science as the *product* of a history [as] a history in progress" (Bensaude-Vincent and Stengers 1996, 4). (McEvoy 2000, 49–50)

This view does not imply that we are not doing philosophy anymore, but, as I have just stated under the premise of analysing the historical framework and the historical settings, we do not claim that the sociology and the philosophy of science are something beyond history.

Kusch has recently presented a fruitful analysis. The opposition to Lavoisier can be categorized as follows (Kusch 2015, 72, citing Chang 2012, 31):

Opponents of Lavoisier's new chemistry

Die-hards	Fence-sitters	New anti-Lavoisierans
J. Hutton, 1726–97	P.-J. Macquer, 1718–84	B. Thomson, 1753–1814
J.-A. De Luc, 1727–1817	H. Cavendish, 1731–1810	G. Smith Gibbes, 1771–1851
A. Baumé, 1728–1804	G.-C. Lichtenberg, 1742–99	T. Thompson, 1773–1852
J. C. Wiegleb, 1732–1800	L. Crell, 1745–1816	J. W. Ritter, 1776–1810
J. Priestley, 1733–1804	C.-L. Berthollet, 1748–1822	H. Davy, 1778–1829
T. Bergman, 1735–84	J. Gadolin, 1760–1852	
J. Watt, 1736–1819	F. Gren, 1760–98	
B.-G. Sage, 1740–1824		

> imply that unifying timeless standards, as for example proposed by Popper, are correct. Standards may change and it is only ex-post plausible or rational how scientists act and think in concrete situations.

Die-hards	Fence-sitters	New anti-Lavoisierans
C. W. Scheele, 1742–86		
J. C. Delamétherie, 1743–1817		
J. B. Lamarck, 1744–1829		
A. Crawford, 1748–95		
J. F. Westrumb, 1751–1819		
R. Harrington, 1751–1837		

Why does this table matter for philosophers? It shows that scientists share a kind of pluralism. It does not show that a strong SSK (sociology of scientific knowledge) programme explains in the best possible manner why giving up the phlogiston theory was so hard. But, is pluralism just another circumscription for irrationalism, as suggested by Hasok Chang? Chang supposes that SSK would deflate "the special authority of science as a whole by reducing the justification of scientific beliefs to social causes" (Chang 2012, 248). I am not a supporter of a strong SSK programme, but a thorough analysis of convictions of scientists shows that they only behave irrationally if there is a dualistic account, like "opt for Lavoisier or stay with the phlogistonists". In the 1780s, scientists had good reasons to pursue more than one theory even if they did not fully recognize the implications of their theories. We will argue in section 2.4.2 how an integrative account can explain some crucial shifts in thinking about phlogiston and Lavoisier's new Bundle Theory.

Thesis 3: In the light of Lavoisier's chemical achievements (1770–1790), the most prominent theses relating to theoretical scientific changes are dubious. If the analysis of knowledge formation within the framework of epistemic networks is important, it would mean that the philosophy of science should care much more about the underlying epistemic processes (see also section 2.4.2).

So, what is wrong with the prominent theses relating to scientific changes? They asserted that the scientists advocating rival paradigms cannot communicate with each other because they do not fully understand the rival theory. But this is not true. Lavoisier communicated with Macquer, Berthollet, Priestley, and Cavendish, who were more or less phlogistonists but also agreed with Lavoisier that quantification and simpler nomenclature in chemistry are some kind of progress.

Second, in an early stage of paradigm change, only a few scientists opt for the new paradigm. This typical behaviour has two effects: it promotes rapid change of acceptance of the new paradigm and then decelerates the acceptance of the new paradigm, as more and more adherers of the old paradigm become aware of the new paradigm and its advantages compared with the old one (Kuhn and Stegmüller). The following Kuhnian theses are also dubious with respect to reconstructions of the Chemical Revolution.

- The change of guiding assumptions for research (shifts of paradigms) is rapid and total
- The whole scientific community is switching to the new theory (the new paradigm)
- Young scientists show a much stronger tendency to get converted into the new paradigm, while older scientists stubbornly stick to the old paradigm (an earlier version of this thesis by Planck, modified by Kuhn and Toulmin).

But does this imply that Kuhn is—taken as a whole—misleading? No. On the contrary, Lavoisier offered, in a nutshell, a kind of Kuhnian account. Lavoisier "articulated a clear conception that in any science two kinds of discoveries arise. The first is cumulative and meaningful in relation to an existing body of knowledge. The second overturns existing systems and leads to revolution *in* the science" (Perrin 1990, 261).

Lavoisier in his own words:

> Among the new facts that continually present themselves to chemists, a large number are worthwhile only to the extent that they are linked to one another, and that they form, in some manner, a continuity, a system. Facts of this sort should be retained attentively by chemists, who ought to ponder them, combine them, link them together by means of other facts, and only make them public after having drawn from them all the possible advantage. *There is another class of experiments that change the entire face of the art, that overturn [systems and establish others at the same time] accredited systems, that open new courses of experiment and reasoning.* One could not be too precipitous in publishing facts of this nature. To keep them from the public would retard the progress of the science. (Lavoisier, quoted in Perrin 1987, 401–2, italics added)

Lavoisier revised this draft of his paper and reformulated the italicized phrase above: "There is another class of experiments that overturn accredited systems, that open new courses of experiment and reasoning, *in a word, that [...] are destined to make revolution in the science*" (Perrin 1987, 402, italics in the original). He could hardly have been more explicit. In fact, he possessed a clearly articulated concept of "progress through revolution" 200 years before Thomas S. Kuhn developed it into an interpretive framework.

2.4.2 Why is it a Chemical Revolution? Lavoisier, Propaganda, and Rational Discourse

Lavoisier, as the key figure of the Chemical Revolution, was, in fact, as we have seen, in some respects, revolutionary. The main thesis that we will elaborate is:

The Chemical Revolution could be understood as a dialectical process of rationalization of experiments and chemical language in combination with new vocational training for chemists and striving for "intellectual empathy" (Chang).

One part of the dialectical process is the oscillating between rational arguments, propaganda, and evaluating chemical knowledge against the background of communication strategies. Our integrative and holistic account of chemical revolution (for a similar view, see McEvoy 2010) assumes that pure philosophy of science is as much inappropriate as any other single discipline since it has to disregard important elements.

To begin with a striking feature of scientific socialization, we refer to McEvoy:

The difference between Lavoisier's corporative view of knowledge and Priestley's individualistic epistemology highlights the difference between the institutional organization of French and British scientific disciplines in the late 18th century. In the highly organized and centralized community of France the pressure of formal education, centralized learned societies, employment opportunities, and a competitive system of reward and recognition meant that aspiring French chemists had little choice but to follow the intellectual lead of the academicians in Paris [...] (McEvoy 1988, 210–11)

From our point of view, those integrated results of social epistemology and the sociology of science and the knowledge-forming processes have to be partly reconstructed by analysing Lavoisier's correspondence. The laboratory logbooks have still not been made accessible to researchers. As we have pointed out, the propaganda instruments and Lavoisier's lecture offer an account that sheds light on how Lavoisier developed his scientific knowledge. In his famous *Traité élémentaire de chimie*, Lavoisier apologized for not giving enough credit to his collaborators and colleagues (see Crosland 2003, 335; Frercks 2008, 316). In his *Traité*, Lavoisier wrote:

If at any time I have adopted, without acknowledgement, the experiments or the opinions of M. Berthollet, M. Fourcroy, M. de Laplace, M. Monge, or in general, of any of those whose principles are the same with my own, it is owing to this circumstance that frequent intercourse, and the habit of communicating our ideas, our observations, and our way of thinking to each other, has established between us a

sort of community of opinions, in which it is often difficult for everyone to know his
own [...] (transl. by Crosland 2003, 335).

Thus, it is not justified to think of the *Traité* as a monolithic system of a single
genius. The "adoption of thoughts of his colleagues" (Frercks 2008, 316) are owed
to, as Lavoisier put it, "the customs of living together" (Lavoisier [1789] 2008, 25).
Furthermore, Lavoisier explicitly wrote about "commonly shared/ownership of
ideas" (Lavoisier [1789] 2008, 25). From the perspective of the sociology of science,
Lavoisier gave us an early modern example of an interdisciplinary research group.
Physicians, chemists, and mathematicians (e. g. Laplace, Lagrange, and Meusnieur de
la Place) collaborated (Perrin 1988b, 116). More tactical reasons motivated Lavoisier
to push his main theoretical insights rather than promote its reception in a more
objective, neutral way.[28] Not the publication of his *Traités*, but the establishment of
the new journal *Annales de chimie* leveraged his comprehensive success.[29]

We will now analyse four main reasons why Lavoisier was, all said and done,
the main figure of a Chemical Revolution (see bullets below).

First, in addition to McEvoy (2010, 255), it was Lavoisier's communitarian,
collaborative view on chemistry that promoted his rationalism compared with
Priestley's rather individualistic empiricism. Such dichotomies are, of course,
dangerous. But Lavoisier's communication of his rationalism strongly suggests
that this rationalism was rather unique and new:

- Rational nomenclature (a project that Lavoisier did not complete, but the rationalization of language is crucial for the communication of chemistry).
- Textbooks should, in Lavoisier's view, revolutionize the vocational training of
  chemists (see Bensaude-Vincent 1990, 435–460). While it took almost a lifetime
  of training without reliable textbooks, one could be a fully skilled chemist in
  a few years. Of course, the tradition of textbooks was almost 200 years old
  when Lavoisier entered the stage, but the professionalization of chemistry in

---

28 It is no exaggeration to classify this as "propaganda" (see Crosland 2009, 93–114, espe-
    cially 99, n. 33).

29 It is beyond the scope of this paper to focus on transnational rivalries, which definitely
    had great influence on personal communication strategies and scientific communica-
    tion styles in general (Gross, Harmon, and Reidy 2002). An example of nationalistic
    preoccupation is the glorification of Lavoisier by Dumas (the editor of Lavoisier's "Traité
    élémentaire de chimie"). Dumas called the *Traité* a "Gospel of Chemists": "One word
    about Lavoisier, who I introduce to you at the moment, as he articulates his 'fiat lux',
    disrupting the mist with his undoubting hand. The old chemistry failed to do so. In this
    moment, I tell you, his powerful voice, the first aurora enlightened the darkness that
    has to fade away in view of the fire of his genius" (quoted in Frercks 2008, 335).

France, in combination with Lavoisier's *Traité*, could be termed as a part of the Chemical Revolution.

- At the ontological-chemical level—that is also part of a vivid communication of the Lavoisier group—we fully agree with Boantza and Gal (2011, 326) who put this part of the revolution in a formula. According to Boantza and Gal, Lavoisier achieved "homogeneous material infrastructure to all chemical phenomena". That is, in fact, a main difference between phlogistonists and Lavoisier, without lowering the contributions of, for example, Cavendish and Priestley to chemistry.

To be more concrete on Lavoisier's view of the abandonment of phlogiston, we will take a closer look at his lecture to the Royal Academy in Paris in 1783 (for an English translation of this lecture, see Best 2015, 137–151; 2016, 3–13). We can have a look at Lavoisier's rationalism in situ, to use a common chemical phrase. Lavoisier began his lecture by referring to deductive methods that yield the oxygen principle as a simple principle. Simplification, in combination with truth (here understood in the framework of the correspondence theory of truth), is, in fact, an element of rationalism:

> I have deduced all explanations from one simple principle: that is, that pure air (vital air) is composed of a special characteristic principle—which forms its base and which I have named the oxygen principle—combined with the matter of fire and heat. Once this principle was accepted, the principal difficulties of chemistry seemed to fade and dissipate and all the phenomena were explained with astonishing simplicity. (Best 2015, 139)

Lavoisier went on with a chemical-philosophical analysis of Stahl's achievements. He gave a lot of credit to the genius of Stahl and to phlogiston theory in general. Later, in his lecture (Best 2015, 143), Lavoisier argued like a critical rationalist, i. e. he thought that "any facts in doubt were rejected at first". He was also aware that new theories have to face some starting problems simply because they are new. Lavoisier, quite self-confident then, stated that the impartial scientific public view shared his theory. Finally, Lavoisier argued that he was "regarded as the true discoverer of the cause of the increase in weight of metallic calces" (Best 2015, 143). A kind of *experimentum crucis* is the following well-known and striking experiment:

> When very pure charcoal is burnt in vital air, all of the charcoal disappears and the vital air is converted into fixed air. If the process is conducted in a closed vessel, weighed precisely before and after the combustion, neither increase nor decrease in weight can be detected. But the air inside the vessel in which combustion takes place, instead of weighing 1.2670 g per litre, weighs 1.86 g. [...]. The increase in absolute

weight of this air is found to be exactly equal to the weight of charcoal that was used. (Best 2015, 146)

Lavoisier concluded in the following logical argument that something has to be wrong with the Stahlian theory that no rational chemist can accept. I have italicized the essential passage:

> If one asks the majority of chemists—partisans of Stahl's doctrine—for an explanation of what happens in this experiment, they will be forced to recognise: Firstly, that it releases the matter of heat and light, which escapes through the vessels and dissipates. Yet, since the weight of the vessels in which one works neither increases nor decreases, they are obliged to admit that the matter of heat and light has no detectable weight. Secondly, they will be forced to recognise that a certain acid—fixed air—forms during combustion. Yet, as the weight of this acid is equal to the combined weight of the vital air and the charcoal, it obviously follows (independently of any system) that a heavy material exists in charcoal that cannot escape through glass vessels and that, consequently, is not the matter of heat and light. *One sees therefore that in the combustion of charcoal Stahl's disciples give the name "phlogiston" to two very different materials—the weightless matter that escapes through the pores of vessels and the weighty matter that unites with vital air to form fixed air.* Therefore, we have two quite distinct substances, which Stahl's disciples conflate: a weightless phlogiston and a weighty phlogiston; one that is the matter of heat, the other that is not. It is by borrowing the properties of one of these substances than the other that they succeed in explaining everything. (Best 2015, 146)

Thus, Lavoisier concluded that phlogiston can explain everything and nothing because of its contradicting characteristics (e.g. it is weightless and it has weight, it is fire, sometimes it is fire "combined with an earthy element" (Best 2015, 150).

Lavoisier's rationalistic agenda for his new chemistry can be found in Best (2016, 3). He claimed a "more rigorous style of reasoning, to scrutinise the facts with which this science enriches itself every day and cast aside what mere reasoning and prejudices add to it. It is time to distinguish fact and observation from what is theoretical and hypothetical" (ibid.).

To sum it up, Lavoisier's rationalism, and especially, the way of communicating this rationalism is striking. The inherent interrelation of Lavoisier's rationalism with the unifying force of his Bundle Theory has still not been analysed in detail. For the Bundle Theory, we refer to Perrin, who listed seven elements of Lavoisier's chemical concept:

> Among the most important, in approximate order of their unveiling, were (i) the absorption of air in combustion and calcinations, (ii) the analysis of atmospheric air into two distinct gases, (iii) the oxygene theory of acids, (iv) the caloric theory of heat

and the vapor state, (v) the composition of water, (vi) the rejection of phlogiston, and (vii) the new nomenclature. Perrin (1988b, 115)

It is beyond the scope of this paper to analyse the relations of these elements. A lot of research has been done on the dissemination of knowledge via the Lavoisier network and via the translation of chemical works (see, e. g., Bret 2016, 122–142; Carneiro, Diogo, and Simões 2006, 671–692). We cannot spend time on this in detail but it is certainly not arbitrary that these networks, the number of translators of Lavoisier's works, and Lavoisier's knowledge of the competing Phlogiston Theory, were more effective than, for example, Priestley's, Cavendish's, and Scheele's networks.[30]

2.5   Conclusion

To summarize, first, the theoretical frameworks of Priestley, Scheele, and Lavoisier were very different. In terms of interpreting the "discovery of oxygen", they had good reasons to embody "oxygen" in their specific frameworks of knowledge. Second, the so-called "Chemical Revolution" is an excellent case study for explaining the formation of scientific communities because Lavoisier did communicate with more than 40 colleagues in a rather modern interdisciplinary style. Social factors like age, nationality, references relating to simplicity or elegance of nomenclature, belonging to the upper class of the super-rich, and the use of propaganda are important when we try to reconstruct what Lavoisier had thought to be true. Rejecting phlogiston theory as a whole is as irrational and fruitless as asserting that Lavoisier's oxygen theory is—in plain and simple terms—correct. Third, we can actually argue in favour of a chemical revolution, but not because of an abrupt

---

30 Finally, as a result of Lavoisier's success, not to say his triumph, the propaganda machine of Lavoisier celebrated his victory over the phlogiston theory in a rather dramatic, or polemical, but surely impressive, play (see Greenberg 2007, 317–18): "Hassenfratz suggested two possibilities for the play. One involved a grand battle. Oxygen's troops included carbonates, phosphates, sulfates, etc., against the allies of Phlogiston, *acidum pingue* and *acide igne*. The other was a confrontation between handsome Oxygen, with his brother-in-arms Hydrogen at his side, and the deformed Phlogiston already missing an arm. At Phlogiston's side is *acidum pingue*, already dead, and *acid igne*, ashen, defeated and dying of fear. Oxygen is poised to lop off Phlogiston's remaining arm. A play was apparently performed and reported to Crell's journal *Chemische Annalen* by a Dr von E**. Phlogiston was placed on trial, weakly defended by Stahl, and then burned at the stake."

change and not because Lavoisier was simply revolutionary in his attitudes and in his chemical views. We have argued that Lavoisier, as the head of a scientific research group, was part of a chemical revolution for several reasons: (1) a revolution in the vocational training for chemists, (2) a growing and intensified importance of manifold forms of communication, (3) the development of a chemical theory with unifying and explanatory power as a "package deal", and (4) a special version of rationalism that was greatly influenced by social factors and shaped a new style of chemical reasoning.

References

Allchin, Douglas. 1997. Rekindling phlogiston: From classroom case study to interdisciplinary relationship. *Science & Education* 6: 473–509.

Ash, Mitchell G. 2012. Wissenschaftsgeschichte und Wissenschaftsphilosophie: Einführende Bemerkungen. *Berichte der Wissenschaftsgeschichte* 35: 87–98.

Boantza, Victor D., and Ofer Gal. 2011. The 'absolute existence' of phlogiston; the losing party's point of view. *British Journal for the History of Science* 44 (3): 317–342.

Benfey, Theodor. 2000. Reflections on the philosophy of chemistry and a rallying call for our discipline. *Foundations of Chemistry* 2: 195–205.

Bensaude-Vincent, Bernadette. 1983. A founder myth in the history of science? The Lavoisier case. In: *Functions and use of disciplinary histories (Sociology of the Sciences. A Yearbook, vol. 7)*, ed. Loren Graham, Wolf Lepenies, and Peter Weingart, 53–178. Dordrecht: Springer.

Bensaude-Vincent, Bernadette. 1990. A view of the chemical revolution through contemporary textbooks: Lavoisier, Fourcroy and Chaptal. *Journal for the History of Science* 23: 435–460.

Bensaude-Vincent, Bernadette. 2014. Philosophy *of* chemistry or philosophy *with* chemistry? *Hyle* 20: 59–76.

Bensaude-Vincent, Bernadette, and Isabelle Stengers. 1996. *A history of chemistry*. Cambridge, MA: Harvard University Press.

Beretta, Marco. 1993. *The enlightenment of matter. The definition of chemistry from Agricola to Lavoisier*. Canton, MA: Science History Publications.

Best, Nicholas W. 2015. Lavoisier's "Reflections on phlogiston" I: Against phlogiston theory. *Foundations of Chemistry* 17: 137–151.

Best, Nicholas W. 2016. Lavoisier's "Reflections on phlogiston" II: On the nature of heat. *Foundations of Chemistry* 18: 3–13.

van Brakel, Jaap. 1997. Chemistry as the science of the transformation of substances. *Synthese* 111: 253–282.

van Brakel, Jaap. 2014. Philosophy of science and philosophy of chemistry. *Hyle* 20: 11–57.

Bret, Patrice. 2016. The letter, the dictionary and the laboratory: Translating chemistry and mineralogy in eighteenth-century France. *Annals of Science* 73 (2): 122–142.

Cahn, Ralph M. 2002. *Historische und Philosophische Aspekte des Periodensystems der Elemente*. Karlsruhe: HYLE Publications.

Carneiro, Ana, Maria Paula Diogo, and Ana Simões. 2006. Communicating the new chemistry in 18th-century Portugal: Seabra's *Elementos de Chimica*. *Science & Education* 15: 671–692.

Carrier, Martin. 2009. Antoine L. Lavoisier und die Chemische Revolution. In *Das bunte Gewand der Theorie: Vierzehn Begegnungen mit philosophierenden Forschern*, ed. Astrid Schwarz and Alfred Nordmann, 12–42. Freiburg: Alber.

Chang, Hasok. 2009. We have never been Whiggish (about Phlogiston). *Centaurus* 51: 239–264.

Chang, Hasok. 2010. The hidden history of phlogiston: How philosophical failure can generate historiographical refinement. *Hyle* 16 (2): 47–79.

Chang, Hasok. 2012. *Is water H_2O? Evidence, realism and pluralism.* Heidelberg: Springer.

Chang, Hasok. 2015. The chemical revolution revisited. *Studies in History and Philosophy of Science* 49: 91–98.

Crosland, Maurice. 1980. Chemistry and the chemical revolution. In *The ferment of knowledge*, ed. George Sebastian Rousseau and Roy Porter, 389–416. Cambridge: Cambridge University Press.

Crosland, Maurice. 1995. Lavoisier, the two French revolutions and "the imperial despotism of oxygen". *Ambix* 42 (2): 101–108. doi:10.1179/amb.1995.42.2.101.

Crosland, Maurice. 2003. Research schools of chemistry from Lavoisier to Wurtz. *British Journal for the History of Science* 36 (3): 333–361.

Crosland, Maurice. 2005. Relationships between the Royal Society and the Académie des Sciences in the late eighteenth century. *Notes & Records of the Royal Society* 59 (1): 25–34.

Crosland, Maurice. 2009. Lavoisier's achievement; more than a chemical revolution. *Ambix* 56 (2): 93–114. doi:10.1179/174582309X441417.

Donovan, Arthur. 1988a. Introduction, in *The chemical revolution: Essays in reinterpretation*, ed. Arthur Donovan, special issue, *Osiris* 4 (2): 5–12.

Donovan, Arthur. 1988b. Lavoisier and the origins of modern chemistry, in *The chemical revolution: Essays in reinterpretation*, ed. Arthur Donovan, special issue, *Osiris* 4 (2): 214–231.

Fors, Hjalmar. 2003. *Mutual favours: The social and scientific practice of eighteenth-century Swedish chemistry.* Skrifter 30. Uppsala: Institutionen för idé- och lärdomshistoria, Uppsala universitet.

Frercks, Jan. 2008. Kommentar. In *System der antiphlogistischen Chemie*, by Antoine L. Lavoisier, trans. Friedrich Hermbstaedt and Jan Frercks, 181–412. Frankfurt am Main: Suhrkamp.

Gale, George. 2003. Scientific explanation. In *The Cambridge history of philosophy 1870–1945*, ed. Thomas Baldwin, 608–620. Cambridge: Cambridge University Press.

Gavroglu, Kostas. 1997. Philosophical issues in the history of science. *Synthese* 111: 283–304.

Giere, Ronald M. 1973. History and philosophy of science: Intimate relationship or marriage of convenience? *British Journal for the Philosophy of Science* 24: 282–297.

Gooday, Graeme. 2008. Placing or replacing the laboratory in the history of science? *Isis* 99: 783–795.

Gough, Jerry B. 1988. Lavoisier and the fulfillment of the Stahlian revolution, in *The chemical revolution: Essays in reinterpretation*, ed. Arthur Donovan, special issue, *Osiris* 4 (2): 15–33.

Greenberg, Arthur. 2007. *From alchemy to chemistry in picture and story.* Hoboken, NJ: Wiley.

Gross, Alan G., Joseph E. Harmon, and Michael S. Reidy. 2002. *Communicating science: The scientific article from the 17th century to the present.* West Lafayette: Parlor Press.

Holleman, Arnold F., Nils Wiberg, and Egon Wiberg. 2007. *Lehrbuch der Anorganischen Chemie.* 102nd ed. Berlin: Springer.

Holmes, Frederic L. 1988. Lavoisier's conceptual passage, in *The chemical revolution: Essays in reinterpretation*, ed. Arthur Donovan, special issue, *Osiris* 4 (2): 82-92.

Holmes, Frederic L. 1997. What was the chemical revolution about? *Bulletin of the History of Chemistry* 20: 1-9.

Hoyningen-Huene, Paul. 2008. Thomas Kuhn and the chemical revolution. *Foundations of Chemistry* 10: 101-115.

Jansen, Walter. 1994. Antoine Laurent Lavoisier. *Chemie Konkret* 1 (3): 162.

Kitcher, Philip. 2011. Epistemology without history is blind. *Erkenntnis* 75: 505-524.

Kim, Mi Gyung 2005. Lavoisier, the father of modern chemistry? In *Lavoisier in perspective: Proceedings of an international symposium*, ed. Marco Beretta, 167-191. München: Deutsches Museum.

Kim, Mi Gyung 2008. The "instrumental" reality of phlogiston. *Hyle* 14: 27-51.

Kusch, Martin. 2015. Scientific pluralism and the chemical revolution. *Studies in History and Philosophy of Science* 49: 69-79.

Labinger, Jay A. and Stephen J. Weininger. 2005. Kontroversen in der Chemie: Wie beweist man ein Negativum? – Die Fälle Phlogiston und Kalte Fusion. *Angewandte Chemie* 117: 1950-1956.

Ladyman, James. 2011. Structural realism versus standard scientific realism: the case of phlogiston and dephlogisticated air. *Synthese* 180: 87-101. doi:10.1007/s11229-009-9607-8.

Lavoisier, Antoine L. (1789) 2008. *System der antiphlogistischen Chemie*. Translated by Friedrich Hermbstaedt and Jan Frercks. Commentary by Jan Frercks. Frankfurt am Main: Suhrkamp.

Lamża, Łukasz. 2010. How much history can chemistry take? *Hyle* 16 (2): 104-120.

McEvoy, John G. 1988. Continuity and discontinuity in the chemical revolution, in *The chemical revolution: Essays in reinterpretation*, ed. Arthur Donovan, special issue, *Osiris* 4 (2): 195-213.

McEvoy, John G. 2000. In search of the chemical revolution: Interpretative strategies in the history of chemistry. *Foundations of Chemistry* 2: 47-73.

McEvoy, John G. 2007. Priestley and Lavoisier. Essay review. *Annals of Science* 64 (4): 595-605.

McEvoy, John G. 2010. *The historiography of the chemical revolution: pattern of interpretation in the history of science*. London: Pickering & Chatto.

Meinel, Christoph. 2000. Chemische Laboratorien: Funktion und Disposition. *Berichte zur Wissenschaftsgeschichte* 23: 287-302.

Melhado, Evan M. 1985. Chemistry, physics, and the chemical revolution. *Isis* 76: 195-211.

Müürsepp, Peeter. 2016. Chemistry as a practical science (Edward Caldin revisited). *Foundations of Chemistry* 18: 213-223.

Nordmann, Alfred. 2002. Die im Lichte sieht man nicht? (Nackte Tatsachen in der Wissenschaftskritik). In *Wissen und soziale Konstruktion in Geschichte, Wissenschaft und Kultur*, ed. Claus Zittel, 47-65. Berlin: Akademie Verlag.

Perrin, Carleton E. 1987. Revolution or reform: The chemical revolution and eighteenth century concepts of scientific change. *History of Science* 25 (4): 395-423.

Perrin, Carleton E. 1988a. Research traditions, Lavoisier, and the chemical revolution, in *The chemical revolution: Essays in reinterpretation*, ed. Arthur Donovan, special issue, *Osiris* 4 (2): 53-81.

Perrin, Carleton E. 1988b. The chemical revolution: Shifts in guiding assumptions. In *Scrutinizing science: Empirical studies of scientific change*, ed. Arthur Donovan, Larry Laudan and Rachel Laudan, 105-124. Dordrecht: Kluwer.

Perrin, Carleton E. 1990. Chemistry as peer of physics: A response to Donovan and Melhado on Lavoisier. *Isis* 81 (2): 259–270.

Priestley, Joseph. 1774ff. *Experiments and observations on different kinds of airs.* Philosophical transactions (6 Vol.). London (vol. 1-4) 1774, 1775, 1777, 1779; Birmingham (vol. 5-6) 1781, 1786 (1790: Complete edition, 3 vol. Birmingham). London: Royal Society of London.

Richardson, Alan. 2008. Scientific philosophy as a topic for history of science. *Isis* 99 (1): 88–96.

Scheele, Carl W. 1777. *Chemische Abhandlung von der Luft und dem Feuer: Nebst einem Vorbericht von Torbern Bergman.* Uppsala: Swederus. http://runeberg.org/scheelch/

Scheele, Carl W. 1780. *Chemical observations and experiments on air and fire.* Trans. Johann Reinhold Forster. Ed. Torbern Bergman, Richard Kirwan and Joseph Priestley. London: Johnson.

Siegfried, Robert. 1988. The chemical revolution in the history of chemistry, in *The chemical revolution: Essays in reinterpretation,* ed. Arthur Donovan, special issue, *Osiris* 4 (2): 34–50.

Stadler, Friedrich. 2012. History and Philosophy of Science: Zwischen Deskription und Konstruktion. *Berichte zur Wissenschaftsgeschichte* 35: 217–238.

Stewart, John. 2012. The reality of phlogiston in Great Britain. *Hyle* 18 (2): 175–194.

Stolz, Rüdiger 1991. Die Chemische Revolution des 18. Jahrhunderts und ihre Wirkung auf das 19. Jahrhundert. *Rostocker Wissenschaftshistorische Manuskripte* 20: 46–50.

Ströker, Elisabeth. 1982. *Theorienwandel in der Wissenschaftsgeschichte: Chemie im 18. Jahrhundert.* Frankfurt am Main: Vittorio Klostermann.

Sukopp, Thomas. 2015. Naturalism in philosophy of chemistry; or: Why metaphysics of nature matters. In *A companion to naturalism,* ed. Juliano do Carmo, 238–255. Pelotas, Brazil: Dissertatio Filosofia.

Thagard, Paul. 1994. Mind, society, and the growth of knowledge. *Philosophy of Science* 61: 629–645. doi:10.1086/289826.

Willeford, Bennett R. 1979. Das Portrait: Joseph Priestley (1733-1804). *Chemie in unserer Zeit* 13 (4): 111–117.

Woodcock, Leslie V. 2005. Phlogiston theory and chemical revolutions. *Bulletin of the History of Chemistry* 30 (2): 63–69.

From Interaction to Integration?
Transnational Scientific Communities in Biology

3

Bertold Schweitzer

Abstract

In the biological sciences, various transnational networks, clubs, and scientific organisations have shaped cooperation and influenced political integration. This chapter presents case studies on early modern naturalists' correspondence, on networks of evolutionary biologists, of the marine station in Naples as a centre of communities of experimentalists, and the foundation of the European Molecular Biology Organisation, suggesting the following insights: First, from the early modern period on, transnational scientific communities of a self-described or implicit European character can be identified, though the sustainability and impact of these communities' Europeanness seems uneven. Second, some of these communities contribute to a European identity and cause spillover effects of varying quality. Third, some developments exemplify the Haasian neofunctionalist logic of "spillovers" and help to explain how processes of scientific community-building can be understood as processes that are both self-reinforcing (leading to intensified scientific cooperation) and motivate further political integration. Still, the shift of scientists' loyalties to a European level seems incomplete at best, since frequently scientists still need incentives in order to be brought to cooperate with partners from other European countries. Fourth, we find that in general, science turns to politics for funds, and politics to science for "innovation" and prestige. Overall, effects of science for promoting European identity exist but are quite limited.

3.1   Introduction

In the context of exploring the role of scientific elites in social and political integration processes in Europe, this chapter examines the development of European transborder communities in the biological sciences, their role in transnational integration in science, and their potential of influencing social and political integration processes beyond the realm of science.

This chapter will examine the contributions of the biological sciences to the establishment of scientific communities across borders, and will explore the influence of scientific integration on social, cultural, and political integration as well as the interactions, spillover effects, and feedback mechanisms at work between scientific, social and political integration.

It will start by exploring a selection of relevant examples of processes of scientific integration across borders in the biological sciences, and will look at prerequisites, mechanisms and effects of both successes and failures of scientific community-building and integration.

Beyond that, it will focus on two further avenues of investigation: First, we will explore the question of how processes of scientific, social and political integration interact with each other, and, more specifically, whether either of these unintendedly lead to or intendedly can be used to promote other forms of integration. Second, it will try to answer the the question of whether the biological sciences in Europe obtained a specifically European character, and, if so, whether this contributed to the emergence of a specifically European scientific identity, and whether this in turn did unfold detectable effects regarding European social and political integration.

As the chapter unfolds, we will explore a number of questions that can be developed from an extension of Ernst B. Haas's classical definition of "political integration" (Haas 1958, 16), among these: Do scientists actually shift their loyalties, expectations and (political) activities to a new (European?) centre? Is this a self-reinforcing process, i. e., does it follow the Haasian neofunctionalist logic of spillovers? Does this contribute to European identity and cohesion?

This chapter seeks to portray some fields and research areas in the biological sciences. It has no aspiration of covering the entire field; in particular it does not attempt to trace transborder community-building efforts in medicine and related fields.

3.2 Case Studies

3.2.1 Early Modern Natural History

The century between around 1520 and 1620 saw a renewed interest in both botany and zoology, and fascination with animals and plants spread throughout Europe (Mayr 1982, 167). Arguably, the first example of the formation of a European scientific community in the biological sciences occurred during this stage of early modern natural history: "At the beginning of the sixteenth century there was no European community of expert natural historians; by the end of that century such a community did exist." (Egmond 2007, 160)

Two of the most important figures in this field were the zoologist Conrad Gesner (1516–1565), notable for his 4000-page encyclopedia *Historia Animalium*, and the botanist Carolus Clusius (1526–1609), who published, amongst others, studies on the Spanish and Austrian floras. Both maintained a vast network of correspondence: Clusius, for example, exchanged letters on natural history with more than 300 people throughout Europe, of which around 1500 have been preserved (see also Mayr 1982, 167; Egmond 2007, 166–67). This correspondence was not only used to exchange information and discuss research findings, just like, for example, Charles Darwin would still do around three centuries later. The exchange of letters, publications, manuscripts, and biological samples also served to negotiate acceptable scientific practices—the "shared notions of validity" of epistemic communities—, and establish natural history alongside more traditional disciplines such as medicine as a new field of expertise.

Moreover, this scientific correspondence began to connect individuals and local groups on a larger scale throughout Europe, and thereby initiated what, by the late Enlightenment, had fully emerged as "some sort of protocommunity (or better, protocommunities) of natural philosophers and natural historians" (Cahan 2003, 295–96). Still, it is in the sixteenth century where we find the first example of community-building processes that have been described as "the formation of a European community of scholars" (Egmond 2007, 159). The vast body of correspondence between natural historians in particular has been described as "the material evidence of the growth of a European community of naturalists", with some of Clusius' correspondents explicitly calling it "a growing community of 'friends' and a republic of letters" (Egmond 2007, 167). Indeed, the "theme of friendship is explicitly mentioned in almost every single letter in the Clusius correspondence" (Egmond 2007, 171), and this notion of friendship seems relevant for the development of science since it contributed to the establishment of acceptable scientific practices beyond the core of "shared notions of validity":

Once a person was recognized as 'a friend'—that is, a fellow member of the network of naturalists—it was not done to withhold information or lie about it, refuse to give counter gifts, be stingy, steal bulbs (or have them stolen by servants), publish someone else's results or discoveries without any form of recognition, bribe agents or brokers in order to obtain rare naturalia that were destined for someone else, plagiarize (although the definition and the notion of authorship were not identical with modern ones), etc. (Egmond 2007, 173–74)

Another influential notion that appears in the correspondence is the idea of a "republic of letters", a term used increasingly from the fifteenth century on. For the group of natural historians, this has been seen as indicating another facet of their group identity: "The fact that a few of Clusius' correspondents explicitly used the term *republic of letters* reflects their self-awareness as a community." (Egmond 2007, 176)

However, it can also be interpreted as expressing a sense of belonging to a wider scientific and literary circle that transcends national boundaries: The French philosopher Voltaire, more than a century later, identified the "republic of letters" as a European phenomenon, where in spite of war, and in spite of religious differences, a republic of the sciences and the arts had established itself:

"On a vu une république littéraire établie insensiblement dans l'Europe malgré les guerres, et malgré les religions différentes. Toutes les sciences, tous les arts ont reçu ainsi des secours mutuels; les académies ont formé cette république." (Voltaire 2016, ch. 34)

Furthermore, Voltaire has been interpreted as arguing in favour of a view that "in spite of political fragmentations, Europeans share the same religious background and the same civil principles" (Bekemans 2014, 25 – but note that Voltaire himself took the republic of letters to have been formed *despite* religious and other differences), and that such shared principles not only "create close ties among nations" but also provide "the decisive element that made Europe the most civilised continent in the world." (Bekemans 2014, 25) The latter view, however, might be regarded as problematic in at least two aspects. The first one, the issue of how this relates to wider social integration, is relativized in the same passage, which points out that "that the feeling of belonging to the same cultural community was shared only by the closed circles of intellectuals with the same classical education, without much influence on ordinary people; for them, the privileged reference was their closer local community and the idea of Europe was seen as abstract." (Bekemans 2014, 25) Still, it might be useful to further investigate the question of whether there were not subtle effects towards a shift of the frame of reference after all. The second issue is to what extent the alleged close ties in the sciences and the arts translate into tighter political integration, something that clearly was not the case yet during the times of either Gesner, Clusius, or Voltaire.

Now, if we accept that during the sixteenth century the first (proto)community of researchers in the biological sciences began to emerge, the questions remains of what, if anything about this community can be identified as specifically European, apart from the manifest fact that, geographically, the network of correspondence, the distribution of books and other publications, and for some individuals, their travels, covered the whole of Europe, and whether these developments were in any way coupled with the development of a European identity or image? The historical analysis on Clusius and other early modern natural historians quoted above asks similar questions: "Was there anything specifically European about this? And can it tell us something about the development of a European identity or image?" (Egmond 2007, 176), but identifies only what amounts to three tentative answers.

The first conjecture is that there might be "a connection between the 'botanical renaissance of the sixteenth century' and the 'scientific revolution of the seventeenth century'" though this connection "remains to be seriously investigated". (Egmond 2007, 176) Still, even when disregarding that fact that the bulk of activity associated with the scientific revolution was done in astronomy and physics, the scientific achievements in the field of biology, as exemplified by the work of William Harvey, Herman Boerhaave, and others, took place in anatomy and physiology, so any connection with botany would seem indirect at best. In addition, characterizing the scientific revolution as "an eminent feature of European science" apparently suggests ascribing a specific European character to it (Egmond 2007, 176). In case we can accept this, the conjecture continues, the community of natural historians, if it can be assumed to have contributed to the scientific revolution in general, might also have contributed more specifically to its Europeanness. Still, the very question of whether there is anything such as a European science, or whether science in general can be said to have a European character, are controversial, to say the least.

The second and third tentative answers focus more directly on the sixteenth-century communities. Here, "we can already isolate two of the specifically European aspects of this community of exchange" (Egmond 2007, 177): The second answer claims that the community of natural historians contributed to establishing Europe as a more concrete entity: "We may argue, therefore, that this 'virtual' community of natural historians in itself helped to give body to Europe—both by its very existence and the frequent exchanges, and by its contacts with parts of the world outside Europe." (Egmond 2007, 177) This claim, though plausible, still appears to be rather elusive, in the sense of not fully specifying what this embodiment amounts to and what concrete effects it is capable of unfolding when it comes to either social or political integration.

The third tenative result addresses styles of conduct and scientific practices, drawing a connection between "European" styles of interpersonal relations and the assumed European character of a community that embraces these styles: "In all of these exchanges the *concept* of Europe seems to have remained completely implicit. Yet, the long-standing European *styles* of honour, friendship and gift exchange that are reflected in the [...] correspondence deeply influenced the practices of this community of scholars. In that indirect way they confirmed and strengthened its European character and probably contributed to the formation of Europe as a concept and mental community." (Egmond 2007, 177) This notion of European styles is concretized by the claim that "new styles of exchange were moulded in the interaction of European experts from different countries, thus creating 'European' styles of scientific exchange and a 'European' community of naturalists." (Egmond 2007, 161) Indeed, this developing style and the associated scientific practices and standards that emerged probably were the most palpable and most consequential effects that can be ascribed to the communication between and communities of European natural historians in the sixteenth century.

To summarize, the aims of early modern naturalist were focused on obtaining information and research materials, and on acquiring biological knowledge and personal prestige within the community. The network was connected by correspondence, the exchange of publications and biological samples, and personal visits, and did not feature any major centre, except for highly regarded individual correspondents, such as Gesner and Clusius. The community's European character and identity is made explicit by some of the correspondents, though any lasting effects appear subtle at best.

3.2.2 Evolutionary Biology in the Nineteenth Century

The following example contrasts various related developments and attempts at community-building in the emerging field of evolutionary biology starting in the middle of the 19th century. Before 1800, many natural historians, for example, Buffon, "had teetered on the brink of evolutionism [...], but none of them had made the decisive step of converting the unbroken chain of a created sequence of ever greater perfection into a line of descent." (Mayr 1982, 343) Lamarck, in a lecture given in 1800, made this "decisive step", and proposed an evolutionary theory, the gradual change and increasing "perfection" or organisms over time, though the physiological and genetic mechanisms responsible that were proposed by him—the inheritance of acquired characters—turned out to be misguided (Mayr 1982, 344). In 1809, Lamarck published his *Philosophie zoologique*—which has also been credited with introducing the term "biology" itself—, and from then on, "no one discussing species, faunas, distributions, fossils, extinction, or any other aspect of organic diversity could afford any longer to ignore the possibility of evolution." (Mayr 1982, 389) However, the details and mechanisms needed to be worked out yet, and during the first half of the nineteenth century, many biologists were developing ideas and collecting relevant evidence.

Following the painstaking collection of empirical support, Charles Darwin eventually came forward with his theory of evolution by natural selection in his *Origin of species*, published in 1859. Darwin's theory of evolution was almost immediately recognized as a major breakthrough in evolutionary biology, and accepted by many influential biologists, thus creating a scientific community of evolutionists. Still, considerable resistance against the revolutionary ideas of evolutionism—both within science and the wider public—needed to be overcome. As a consequence, more tightly-knit scientific communities were formed, dedicating themselves to defending Darwinism against its critics and promoting its acceptance in universities and society.

The purpose of this section is twofold: First, to discuss the community-building in research, and second, to examine efforts to establish networks for promoting the new ideas on evolution.

The research process in evolutionary biology relied on piecing together a wealth of highly diverse evidence, and indeed most researchers, including Darwin, relied on a network of correspondents and informants, not unlike earlier such networks in natural history. Charles Darwin himself was noted for his widespread network of correspondents. Notable European correspondents include Alphonse de Candolle, Franciscus Donders, Anton Dohrn, Johan Georg Forchhammer, Jean Albert Gaudry, Julius von Haast, Ernst Haeckel, Fritz Müller, Melchior Neumayr, Armand de Quatrefages, Japetus Steenstrup, and August Weismann. The "Darwin Correspondence Project" at Cambridge University alone owns 9,000 letters plus copies of another 6,000.

As to the character of Darwin's interactions with his correspondents, some scholars saw evidence that "Darwin used his vast correspondence network just like Clusius—in order to obtain information from a wide geographic and social range of informants—but he usually did not reciprocate, either with ideas or gifts" (see Egmond 2007, 177). Other scholars have contradicted this claim, and have described the relations between Darwin and many of his correspondents as establishing a genuine scientific community: "This sense of a community of endeavour, and of mutual responsibility, was then often acknowledged and developed further through correspondence." (White 2008, 57)

For example, the co-adaptation of flowers and insects was but one research area suggested by Darwin's work that was "immediately taken up by this circle of German botanists, forming a web of mutual exchange and support, centred on discussions of experimental practice." (White 2008, 64) Again, this group of German botanists is explicitly described as a scientific community: "We might imagine such a community as a 'thought' or 'style collective' in the sense proposed by Ludwik Fleck, applying the term broadly as Fleck did, to include economies of scientific work (Fleck 1935)." (White 2008, 64)

However, contrary to what one may have expected, the focus of the interaction in such communities seems to have been less the wish to promote evolutionism in general, and Darwin's theoretical ideas in particular, and more the question of how Darwin's theoretical framework could be made fruitful for suggesting new research questions and new methods for carrying out empirical research in biology:

> Darwin's correspondents saw him as opening new ways of conducting natural history, or as engaged in similar experimental practices and as therefore offering some validation of their own. Their letters formed part of a circulation of specimens and observations, experimental design, mutual citation, translation and promotion. Their correspondence shows that reception is also about these practical connections, about transformations in natural historical practice as much, or more, than natural historical theory. (White 2008, 64)

The second area to be examined is the development of networks for promoting (and defending) Darwin's theories of evolution. Interestingly enough, the major initiatives for establishing such communities seem to have taken place on a national level. Though Darwin's ideas were received with considerable interest throughout Europe, we fail to find evidence for a (pan-)European community of Darwinists. One reason is obviously the increased nationalism in the nineteenth century, an important other one the diverse and often incompatible versions of evolutionism held by individuals or groups on a local or national level, leading to a "growing split among the evolutionists" from around the time of Darwin's death in 1882 (Mayr 1982, 540). In particular, the role of natural selection remained controversial, with Darwin himself having complicated matters, since he famously claimed "[...] I am convinced that Natural Selection has been the main but not exclusive means of modification" (Darwin 1859, 6), and also modified some his views between the various editions of the *Origin of Species*, most notably between the fifth and sixth edition, 1869 and 1872. The controversy about the role of natural selection between broadly Lamarckian views that accepted the inheritance of acquired characteristics, and the panselectionist account of August Weismann asserting natural selection as the *only* mechanism capable of creating *adaptive* features of organisms continued for decades and led to numerous disputes between various factions of self-described Darwinians.

A further indication that the spread of Darwinism was a European phenomenon only in the sense that it happened in many European nations at the same time rather than being connected with a (however virtual) pan-European community of evolutionary biologists is the fact that out of 29 essays in a recently published collection on the reception of Charles Darwin in Europe, 27 discuss the situation in individual states, and none Europe as a whole (Engels and Glick 2008).

What can be identified, however, are communities and networks on a national level. The probably most famous such network that was created for promoting science in general, and Darwin's ideas in particular, was the "X Club" in Britain. For the most part, social and professional networks in 19th-century British science were centered around various learned societies (including the Linnean Society and others), the British Association, and individual universities. One the other hand, there were also small, informal, but influential networks of individuals. These included, for example, the "Philosophical Club", established 1897, which was seeking reform within the Royal Society, and the "B-Club", founded 1865 in order to promote chemistry within the British Association.

For evolutionary biology, the most important association was the "X-Club", founded 1864, which has been described as "the most powerful and influential scientific coterie in England" (Fiske 1939, 144). At its core, the "X-Club" was a dining club of nine men who supported natural selection and academic liberalism in late 19th-century England. Thomas Henry Huxley called the first meeting for 3 November 1864 in London; the club met until 1893. Its members, George Busk, Edward Frankland, Thomas Archer Hirst, Joseph Dalton Hooker, Thomas Henry Huxley, John Lubbock, Herbert Spencer, William Spottiswoode, and John Tyndall already knew each other well. They were united by a 'devotion to science, pure and free, untrammelled by religious dogmas.' (Hooker to Darwin 1865, quoted in MacLeod 1970, 311). After Darwin's *Origin of Species* was published in 1859, the club began working to aid the cause for naturalism and natural history. Its members also backed the liberal Anglican movement of the early 1860s.

The X-Club's key aims included scientific discussion free from theological influence, as well as reforming the Royal Society, and making the practice of science professional. During the 1870s and 1880s, its members became prominent in the scientific community (both within the Royal Society and elsewhere). The X-Club members were seen to have much power in shaping the scientific scene in Britain, and played a significant part in nominating Darwin for the Copley Medal in 1864 (Barton 1998).

Though no direct counterpart to the X-Club seems to have existed in Germany, German evolutionists were facing similar struggles against orthodoxy, here in particular on the question of whether evolutionism should be taught in schools.

After initial local attempts to introduce evolutionary ideas into the curriculum, the teachings of Darwin and Haeckel were banned in Prussia (Storch et al. 2013, 35). Paradoxically, this seems to even have helped the spread of evolutionism, in particular as popularized by Ernst Haeckel in his numerous books (see Richards 2008): "A German pupil remembered: 'Who of us did then not carry next to the official New Testament an unofficial volume of Haeckel?'" (Sommerey 2014).

In conclusion, the X-Club, and other, less tightly-knit communities in, for example, Germany, can be seen as a scientific communities with (scientific *and*) political aims, some of which they were successful in carrying through. Geographically, however, these communities intended to promote an evolutionist worldview were mostly limited to individual nation states.

Given the fact that many evolutionists shared a naturalist worldview, and supported a stronger role of the natural sciences against orthodoxy and institutionalised religion, in particular in schools, universities and education in general, we would have expected them to at least attempt to join forces across borders. Currently, however, there seems to be little evidence for any of this.

In conclusion, the aims of evolutionary biologists after Darwin were, again, focused on gaining access to information and research materials, and on acquiring biological knowledge and personal prestige within the community. In addition, science policy, in particular the promotion of a scientific worldview and the fight against opposing traditional views, in universities and in wider society, gained prominence. The evolutionists' networks were still connected mainly by correspondence, the distribution of publications and biological samples, and personal meetings, with scientific conferences as a platform only gradually. Interestingly, the national networks and clubs seem to have been much more tightly connected and uniform in their aims than any of their transnational counterparts. Some major centres such as London and Paris emerged, where small clubs of personally acquainted individuals became important. Though communities were active in most European countries, no comparable pan-European communities seem to have emerged. The lasting effects on science policy, and teaching and research in universities throughout Europe are undeniable, however.

3.2.3   The "Stazione Zoologica" in Naples, Italy

Also connected with Darwinism and with attempts to promote research in evolutionary biology, the following example explores a much more concrete case of institution and community building: The foundation of the "Stazione Zoologica" in Naples, Italy, by the German zoologist Anton Dohrn (1840–1901). In Europe, interest in sea life and marine organisms had been on the rise during the ninteenth century (Groeben 1985, 7), motivated by a combination of the expectation that the study of sea life—where life was supposed to have originated—would contribute to the understanding of fundamental biological and evolutionary processes, and the opportunities marine organisms provided for observational and experimental studies in morphology and physiology: sea urchins and their eggs, for example, are a traditional model organism in developmental biology. Hence, both marine biologists and experimental physiologists working with marine organisms required easy access to the sea, but this met with many obstacles.

Dohrn was, on the one hand, a zoologist who had himself experienced the practical difficulties of obtaining and preserving, and experimenting with marine organisms, for example, on his research trips to Heligoland, Scotland and Sicily. On the other hand, in 1862 he had become acquainted with Darwin's ideas, and had started to see zoology, and in particular morphology, comparative anatomy, and comparative embryology as tools for clarifying the phylogenetic relationships between organisms, a connection also motivated by Haeckel's claim that the embryonic development of an organism mirrored, or recapitulated this organism's evolutionary history.

Using his own funds, support from various European scientists—Thomas H. Huxley, Charles Darwin, Francis Balfour, Charles Lyell, and others—and the German government, and a construction site provided for free by the city of Naples, Dohrn in 1872 founded the "Stazione Zoologica"—still in existence, and today named "Stazione Zoologica Anton Dohrn". The station was conceived as a research institute devoted to interdisciplinary basic research in biology, and Dohrn's basic idea was to establish an international scientific community provided with laboratory space, equipment, research material, and a library. The station's running costs were paid from income from the "table", or "bench" system—where a university would rent one of the station's laboratory benches one a permanent basis, and be entitled to send one at a time of their staff who would be provided with working space and access to all of the station's facilities—, the sale of scientific journals and specimens, and the income from the station's public aquarium.

From its start, the station proved to be hugely successful: by 1890, 36 scientists from 15 countries worked in the station per year (Groeben 1985, 11), and until 1909, more than 2,200 scientists from Europe had visited the station, and more than 50 benches had been rented out per year (Groeben 2006, 295). In terms of results, Hans Driesch claimed in 1909 that at the station, "9/10 of all basic work in modern zoology has been done" (Groeben 1985, 5).

In terms of community building and the European dimension, the following considerations seem important: First, Dohrn clearly "wanted to make a contribution to advancing Darwinism" (Groeben 1985, 8), thus facilitating the cooperation with and support by British biologists such as Darwin himself and Huxley as well as Darwinists throughout the rest of Europe. Second, however, on a practical level Dohrn was willing to support any kind of research in biology at the station: "Dohrn's concept included strict neutrality with regard to the studies performed by scientists using his facilities, despite his own strong opinions ..." (Engels and Glick 2008, 383). Third, the station provided a centre for European Darwinists, though, as discussed above, this community was internally fractured. For example, Dohrn himself had ended his friendship with Haeckel, the most important popularizer of Darwin's ideas, in the late 1860s when he "could no longer subscribe to Haeckel's philosophical generalizations" (Groeben 1985, 8). Indeed, such controversies seem to have influenced the composition of the group of scientists working at the station: "As a consequence, Haeckel never made use of the Stazione, though some of his students and supporters did. And yet, some degree of rivalry developed between what Maggi defined as the 'Dohrnian current' and the followers of Haeckel" (Engels and Glick 2008, 383). Fourth, and possibly even more importantly, the station became a hub for methodological innovation in biology. In the words of Edmund B. Wilson, "the station has now become practically the headquarters from which most of the leading European laboratories derive their best methods, and where, indeed, much of their most telling work is done" (E. B. Wilson, letter of 9 March 1883, quoted in Fantini 2004, 526). Hence, it created a European community of methodologically like-minded biologists. On a more general level, this illustrates how various scientific communities may coexist: A biologist could, at the same time, be a Darwinist, carry out research on radiolarians, and use microscopic techniques, and hence be a member of three different communities, whose ties may or may not be mutually reinforcing.

In conclusion, this case demonstrates how a scientific community can be brought into being by the planned creation of a suitable hub such as a research facility. As an "an able manager of science" (Groeben 2006, 295), Dohrn created a centre, and thus indirectly helped create a community: The station initiated the formation of a (predominantly) European community of experimental biologists, triggered by the idea and initiative of one scientific manager or entrepreneur. "Dohrn's greatness consisted in creating an organism, the potential of which was larger than national, contemporary and socio-political ties" (Groeben 1985, 22), and his project indeed not only created a mostly European community of experimental and evolutionary biologists, but also motivated the establishment of further biological stations throughout Europe and beyond.

3.2.4 European Molecular Biology Organization

A fourth example, the founding of the European Molecular Biology Organization (including the European Molecular Biology Conference and the European Molecular Biology Laboratory), again highlights a case of community-building mainly from within science, in this case initiated by leading scientists assuming the roles of scientific entrepreneurs or managers of communities.

While the seminal ideas on using molecular approaches in biology can be traced back to the 1930s, molecular biology as a discipline did not exist before the 1950s (see Cairns et al. 2007). Established disciplines included crystallography, microbiology, biochemistry, biophysics, physical chemistry, embryology, and genetics, but these had to be integrated by formulating the specific epistemic aims of molecular biology. These can be characterized as attempting to explain biological functions in terms of (macro-)molecular structures: "the term molecular biology was created to indicate that its subject is to explain the biological function of certain macromolecules by the knowledge of their structure at the molecular level" (Tissières and Kellenberger 1962, quoted in Strasser 2002, 526).

The biological insights gained during the 1950s did much to demonstrate the feasibility of this approach. These insights included the elucidation of the structure of DNA by Watson and Crick in 1953, the demonstration that genes uniquely specify the amino acid sequences of proteins (and that point mutations in genes cause single amino acid changes in proteins, e. g., sickle cell anaemia), and the confirmation that the amino acid sequence of a protein alone determines the three-dimensional structure of proteins, and as a consequence, their function as well.

However, existing organizational structures, in particular the organization of university departments along the lines of traditional discipline boundaries initially hampered the establishment of a new discipline. As a consequence, dedicated research facilities for molecular biology were first established in research institutions rather than universities, and many were using funding from centralised agencies, thus bypassing traditional university structures and rivalries (Strasser 2002, 528). At the outset, new national, and nationally funded institutes for molecular biology were created in Switzerland, Germany, France and Britain, whereas in Belgium and Italy, these were partially funded by EURATOM.

Only with both a dedicated epistemic focus and a number of organizational centres in place it can be said that a community of molecular biologists had formed, and hence it seems fair to conclude that molecular biology "acquired a social reality only around 1960" (Strasser 2002, 516).

National infrastructures and national funding efforts however appeared to be inadequate, in particular when compared to the level of support the USA provided for molecular biology—but also when looking at the opportunities other transnational European projects were providing for fields such as nuclear research.

A rough outline of the developments that led to the foundation of EMBO runs as follows (see Tooze 1986; Strasser 2002; Strasser 2003; Ferry 2014): In December 1962, the molecular biologists Kendrew, Watson, Szilard developed the idea to emulate the existing European Organization for Nuclear Research (CERN), and to establish a European Molecular Biology Organization. In September 1963, a meeting of molecular biologists in Ravello, Italy, decided that a European organization was more realistic than a global one. This led, in 1964, to the establishment of the European Molecular Biology Organization (EMBO) as a non-profit organization registered in Switzerland which was able to secure funds from private donors in order to provide short- and long-term fellowships, and organize courses and workshops. In 1969, "13 European governments were reaching the decision to establish a full-fledged intergovernmental organization, the European Molecular Biology Conference (EMBC), to provide substantial long-term governmental support for EMBO's program" (Tooze 1986, S39). One of the fundamental questions was whether EMBO should focus on the establishment of a central laboratory or whether it should organize a network of research centres; ultimately, EMBO was set up to include elements of both: "EMBC's founding document gave it a dual aim: to provide a secure source of funds for a generic programme that included initially the EMBO fellowships, courses, workshops and administration; and to provide a framework to establish the European Molecular Biology Laboratory (EMBL)" (Ferry 2014, 43). The European Molecular Biology Laboratory (EMBL) was created in 1974, with its main campus located in Heidelberg, Germany.

For explaining the success of EMBO, various reasons have been suggested. One is that the promoters put forward the following arguments in favour of a (European) molecular biology, which proved to be convincing to European governments and other donors: (1) Molecular biology is a "modern" science, using methods from physics, in particular nuclear physics, and takes part in a general trend towards using "atoms for peace", (2) molecular biology carries out fundamental research, but with highly promising medical applications, (3) the USA are much more advanced in science than Europe, but it European molecular biology should be able to help reduce this gap, (4) molecular biology is fundamentally interdisciplinary, and efficient research relies on the division of labour, as in other large-scale projects, and (5) molecular biology transcends the different existing approaches or paradigms in biological science and can hence be expected to be highly objective (Strasser 2002, 530–541).

Another reason for EMBO's success can be seen in the fact that is was a (Western) European project (unlike other initiatives of a global, Atlantic, pan-European, or regional character). Specifically, the EMBO projects were perceived to bolster Europe scientifically and to address the scientific and technological 'gap' between Europe and the USA.

> "EMBO's founding fathers believed that it was essential that these complementary activities [central laboratory, exchange fellowships, training courses, research grants] be developed at the European, international level to ensure that the pool of talent and resources in Europe was most efficiently and effectively mobilized. The hindrances of national boundaries and the restrictions to which national funds and resources are often subject had to be circumvented if Europe as a whole was not to fall further behind the United States." (Tooze 1986, S38)

On a more general outlook, supporting EMBO was seen by participating states as a welcome justification to cooperate in the first place; it this sense EMBO was seen as an opportunity for establishing cooperation in one highly specific area of common interest in order to strengthen political ties in general. This can be seen as one variation of a general neofunctionalist strategy, often referred to as "Schuman doctrine". "Not coincidentally, the importance of science as a bonding agent within Europe paralleled the efforts to tie together the economic future of Europe [...]" (Wang 2015, 149).

To summarize, the aims of molecular biologists in the twentieth century were focused on large-scale interdisciplinary research in a relatively large but distributed community, and on securing funding for costly equipment, a central laboratory, and individual fellowships. Arguments on Europe losing the competition between itself and the US, and on potential medical benefits may have been more instrumental than intrinsic. The community was institutionalised, backed by a formal international organization, and a European network supplemented by a central laboratory. The community's European character has been explicit since EMBO's foundation, and has been shaping identities as well as serving as a nucleus for further steps towards integration of biological sciences in a European context.

3.3   Conclusions

Based on the examples and case studies presented here, a number of tentative conclusions on the role of scientific communities and epistemic networks in the biological sciences on a European level can be drawn:

First, the history of biology shows that there are indeed communities that define themselves as European, while others remain predominantly national networks.

However, the scope, the effects (both short- and long-term), and the sustainability of such transnational communities are highly divergent.

Second, these communities' contributions to a European identity, and spillover effects regarding social, cultural, political integration range from plausible but elusive (as in the case of early modern natural history) to almost textbook-like (as in the case of the European Molecular Biology Organization).

Third, concerning the question of whether scientists, by establishing transborder scientific comunities, shift their loyalties, expectations and activities to a new—for example, European—centre, we do indeed find some evidence for the occurrence of such shifts. However, they usually occur less because scientists define themselves primarily as European, or value European or international cooperation per se (many cooperate internationally anyway), but much more frequently because Europe provides (or provided in some historical context) unique opportunities for carrying out research.

Fourth, as to the question of whether scientific community-building on a European level is a self-reinforcing process, i.e., whether it follows the Haasian neofunctionalist logic of "spillovers", a mixed answer is the best we can give: In part, such processes do seem to follow a neofunctionalist logic, insofar as successfully established high-profile European scientific institutions or organisations do serve as points of reference for either dealing with newly emerging issues, or for organising further scientific disciplines (as in the case of the European Organization for Nuclear Research (CERN) serving as a partial blueprint for the European Molecular Biology Organization). However, the neofunctionalist logic does not seem to apply throughout, as demonstrated by the fact that scientists' preferences, loyalties and activities do not seem to have shifted to a European level in a substantial and sustainable fashion. This view is evidenced by the fact that the majority of EU-funded programmes for scientific research still see the need to considerably nudge scientists towards cooperating with partners from other European countries (see Ytreland 2009), something that would not be expected if scientists' loyalties had indeed shifted thoroughly.

Fifth, as to the issue of scientific communities' contributions to European identity and cohesion, it can be affirmed that in general, scientific success does provide prestige to polities, and Europe and the EU are no exception. The general pattern apparently is that science is turning to politics for funds, and politics is turning to science for "innovation" and prestige. Official voices routinely try to claim enhanced European identity and cohesion amongst the outcomes, but notwithstanding some plausibility, the overall effects are most likely rather limited.

References

Barton, Ruth. 1998. "Huxley, Lubbock, and half a dozen others": Professionals and gentlemen in the formation of the X Club, 1851–1864. *Isis* 89: 410–444. doi:10.1086/384072.

Bekemans, Léonce. 2014. The idea of Europe: Identity-building from a historical perspective. In *Transfigurations of the European identity*, ed. Bulcsu Bognár and Zsolt Almási, 21–41. Newcastle upon Tyne: Cambridge Scholars Publishing.

Cahan, David. 2003. Institutions and communities. In *From natural philosophy to the sciences: Writing the history of nineteenth-century science*, ed. David Cahan, 291–328. Chicago: Univ. of Chicago Press.

Cairns, John, Gunther S. Stent, and James D. Watson, ed. 2007. *Phage and the origins of molecular biology*. Centennial ed. Cold Spring Harbor, NY: Cold Spring Harbor Laboratory Press.

Darwin, Charles. 1859. *On the origin of species by means of natural selection: Or the preservation of favoured races in the struggle for life*. London: John Murray.

Egmond, Florike. 2007. A European community of scholars: exchange and friendship among early modern natural historians. In *Finding Europe: Discourses on margins, communities, images ca. 13th – ca. 18th centuries*, ed. Anthony Molho and Diogo Ramada Curto, 159–183. New York: Berghahn Books.

Engels, Eve-Marie, and Thomas F. Glick, ed. 2008. *The reception of Charles Darwin in Europe*. 2 vols. London: Continuum.

Fantini, Bernardino. 2004. The "Stazione Zoologica Anton Dohrn" and the history of embryology. *International Journal of Developmental Biology* 44: 523–535.

Ferry, Georgina. 2014. *EMBO in perspective: A half-century in the life sciences*. Heidelberg: European Molecular Biology Organization.

Fiske, John. 1939. *The personal letters of John Fiske*. Cedar Rapids, IA: Torch Press.

Groeben, Christiane. 1985. Anton Dohrn – the statesman of Darwinism: To commemorate the 75th anniversary of the death of Anton Dohrn. *Biological Bulletin* 168: 4–25. doi:10.2307/1541316.

Groeben, Christiane. 2006. The Stazione Zoologica Anton Dohrn as a place for the circulation of scientific ideas: Vision and management. In *Information for responsible fisheries: libraries as mediators*, ed. Kristen L. Anderson and Cecile Thiery. Fort Pierce, FL: International Association of Marine Science Libraries and Information Centers.

Haas, Ernst B. 1958. *The uniting of Europe: Political, social and economic forces 1950–1957*. London: Stevens.

MacLeod, Roy M. 1970. The X-Club: A social network of science in late-Victorian England. *Notes and Records of the Royal Society of London* 24: 305–322.

Mayr, Ernst. 1982. *The growth of biological thought: Diversity, evolution, and inheritance*. Cambridge, MA: Belknap Press.

Richards, Robert J. 2008. *The tragic sense of life: Ernst Haeckel and the struggle over evolutionary thought*. Chicago, IL: Univ. of Chicago Press.

Sommerey, Constance. 2014. "Illegal science": The case of Ernst Haeckel (1834–1919) and German biology education. http://www.shellsandpebbles.com/2014/08/04/illegal-science-the-case-of-ernst-haeckel-1834-1919-and-german-biology-education/.

Storch, Volker, Ulrich Welsch, and Michael Wink. 2013. *Evolutionsbiologie*. 3rd ed. Berlin: Springer Spektrum.

Strasser, Bruno J. 2002. Institutionalizing molecular biology in post-war Europe: A comparative study. *Studies in History and Philosophy of Science Part C: Studies in History and Philosophy of Biological and Biomedical Sciences* 33: 515–546. doi:10.1016/S1369-8486(02)00016-X.

Strasser, Bruno J. 2003. The transformation of the biological sciences in post-war Europe. *EMBO Reports* 4: 540–543. doi:10.1038/sj.embor.embor879.

Tooze, John. 1986. The role of European Molecular Biology Organization (EMBO) and European Molecular Biology Conference (EMBC) in European molecular biology (1970–1983). *Perspectives in Biology and Medicine* 29: S38–S46. doi:10.1353/pbm.1986.0017.

Voltaire. 2016. *Siècle de Louis XIV (VI): Chapitres 31–39*. Edited by Diego Venturino. *The Complete Works of Voltaire* 13D. Oxford: Voltaire Foundation.

Wang, Tom C. 2015. Science diplomacy: Transatlantic asset and competition. In *Smarter power: the key to a strategic transatlantic partnership*, ed. Aude Jehan and András Simonyi, 147–154. Washington, DC: Center for Transatlantic Relations.

White, Paul. 2008. Correspondence as a medium of reception and appropriation. In *The reception of Charles Darwin in Europe*, ed. Eve-Marie Engels and Thomas F. Glick, 54–65. London: Continuum.

Ytreland, Ingrid. 2009. *Connecting Europe through research collaborations? A case study of the Norwegian Institute of Public Health*. University of Oslo. https://www.duo.uio.no/handle/10852/17892

Science and the Manufacture of a Political Order

4

A Sociological Study of the European Manned Space Programme

Julie Patarin-Jossec

Abstract

This chapter aims to question the function of science in the sustainment of a UE political power, based on mixed sociological methods. The European contribution to the International Space Station provides some relevant answers. First, I will introduce international collaboration patterns structuring the organization of science aboard the European research facility ("Columbus"), so as to define the context in which the European contribution took place, and the evolution of its space programme, which followed an increasing organizational standardization of space activities. The second section will focus precisely on the consequent organizational requisites for a political stability on the international space stage, highly dependent to the bureaucratic management of spaceflights, which also leads to a form of authoritarian governance of scientific processes. This central place of science in the conditions for political stability supposed that it constitutes axiological foundations for a legitimized occupation of outer space, without any call for strategic and military logics. The last section will then highlight the scientific roots of political mechanisms like nationalism from a critical sociological outlook, where science and its cultural authority facilitate the political integration and rhythm of power relationships in international affairs. In fine, our concern will be to measure how heuristic it would be to make scientific activity a part of the relationships of production which, in every Marxist theory of the State, structures the latter, preventing its reduction to a reified monolith.

4.1   Introduction

During the 2000s, Europe's leaders decided on a strategy for the European Union that would have developed a competitive knowledge-based economy (partly through the creation of a "European Research Area"; see Madsen 2010); science and technology obtained a privileged place in the European integration project. In the case of human spaceflights, or space activities in general, political concern has never been concealed, but science has been presented as the main concern of spaceflights, while military and strategic motivations of occupying outer space lost their credibility. Moreover, an increasing number of experiments (as well as non-scientific activities like the maintenance and the technical improvement of the station), currently performed aboard ISS, are a part of the preparation for future long-duration spaceflights through the solar system. Political concerns are then obvious, but they do not constitute a novelty; presented as an outstanding laboratory, the International Space Station programme was motivated by a diplomatic concern: framing a long-term international cooperation among governments which were mainly in conflict on Earth, when the Cold War was not yet over.

From the anthropology of science, looking for the constitutional order of the research (Brice 2013), to the relations among civil society and science in public policies (Jasanoff 2012; Irwin and Wynne 1996), to the political governance of scientific research (for example: Elzinga and Jamison 1995) and the scientific expertise for political communities (Granjou 2003), or even the coordinating function of scientific instruments among science, state and industry (Joerges and Shinn 2001), the role of science and technology in international relations (Ancarani 1995) and historical relations between science and power (Salomon 1973), the literature is plethoric and various in social science studies. However, the function of science in the state-building process, and in what Nicos Poulantzas defines as the "institutional materiality" of the state (from which it can exercise a political power), remains an analytical tool that has been overlooked.

The European contribution to human space flights has always had to deal with organizational and political difficulties, starting with its internal multiplicity (which does not facilitate a homogenous governance that does not weaken its capacity to exercise a certain power) and its place on the international space stage (between the two opposing powers during the Cold War). This chapter, therefore, aims to determine how the European space agency, through its management of scientific experiments aboard its Columbus module, contributes to maintaining a European state power through the organization of human spaceflights, while one could not talk about an "achieved European state" (see Durand and Keucheyan, 2015).

Consequently, my concern will be to question the function of science in the sustainment of a UE political power, partly considering successive crises in governance since the 1970s. As I will describe below in detail, the European contribution to the International Space Station (1998–2024, hereafter "ISS") provides some relevant answers. First, I will introduce international collaboration patterns structuring the organization of science aboard the European loaded facility, "Columbus", to define the context in which the European contribution took place and the evolution of its space programme, which followed an increasing organizational standardization of space activities. International affairs are highly determined by the management of activities aboard the space station among the infrastructures of different member-states. Ergo, an overview of the genesis of the ISS programme, which is necessary to concretely identify the place of the European space programme. The second section will precisely focus on the consequent organizational requisites for political stability on the international space stage, which are highly dependent on the bureaucratic management of spaceflights, and which also leads to a form of authoritarian governance of scientific processes. UE consequently regulates a political power weakened by what Antonio Gramsci qualifies as an "organic crisis", which is related to an overaccumulation of capital. But if science (the main activity aboard space stations and the ISS) can be useful in political stability, it is because it constitutes axiological foundations for a legitimized occupation of outer space, without a call for strategic and military logics. The last section will then clarify the mechanisms of this overaccumulation crisis and explain how science in manned spaceflights contributed to its resolution, highlighting mechanisms through which the scientific production contributes to this process of political power-building, thanks to its cultural authority and its depoliticized definition. This is because of the values on which it is based in the mainstream history of science (values such as objectivity, axiological neutrality, and universalism). In fine, our concern would be to measure how heuristic it would be to make scientific activity a part of the relationships of production which, in every Marxist theory of the State, structures the latter, preventing its reduction to a reified monolith.

4.1.1   Methods

The following sections are based on empirical data collected as part of a research in sociology that started in October 2015. Three main techniques of investigation were deployed: ethnography at the control centres of the European, French, and Russian space agencies;[1] interviews with scientists, astronauts, policymakers, operators and functionaries of industrials involved in ISS activities;[2] and the consultation of archives at the documentation centres of the Canadian and European space agencies.[3]

Among the five ways to perform scientific experiments under microgravity (sounding rockets, drop towers, photon probes, parabolic flights, and space stations), only the last two are manned, since they imply boarding experimenters. They, consequently, constitute base materials. The ethnographic methods was then applied at the headquarters of space agencies (in Paris, Moscow and Montreal), five control centres (Cologne, Brussels, Oberpfaffenhofen, Toulouse and Moscow), as well as two astronaut training centres (Cologne and Moscow). In particular, the infrastructures of the European space agency were studied, including four of its user support operations centres spread over its member-States (Germany, Belgium, and France). Besides, a case control study allowing multiplicated modalities of experimentations in manned spaceflights was conducted (mixing interviews and observations) during parabolic flights campaigns at Novespace (Merignac, France), where experiments are performed aboard an Airbus A310, reproducing free fall spearheading microgravity. At the opposite of the space station, scientists are experimenters on their own experiments aboard the plane, and consequently do not have to delegate protocols to astronauts.

4.2   Setting the International Context

4.2.1   The European Space Programme: A Brief Genealogy

Created in 1974 in the aftermath of a meeting of 11 countries in Brussels on 31 July 1973, and as a fusion of ELDO (European Launch Development Organization, created in 1972) and ESRO (European Space Research Organization, created in 1962, concomitantly with the inauguration of the Centre National d'Etudes Spatiales), the European space agency was first constituted with Belgium, the UK, Denmark, France, Germany, Italy, Spain, the Netherlands, Sweden and Switzerland. The agency is formed with singular and well-defined purposes: (1) to build a "European

---

1   Respectively: In Cologne (Germany), Oberpfaffenhofen (Germany), Toulouse (France), Korolev (Russia).

2   84 interviews were realised between September 2015 and December 2016.

3   Respectively at the CSA headquarters in Saint-Hubert (Quebec, Canada) and at CADMOS (Toulouse Space Centre, France).

launcher" (the L3S), (2) to contribute to the construction of the American Spacelab, and (3) to construct MAROTS (the Maritime Orbital Test Satellite). Because L3S was mainly led by France, Spacelab by Germany, and MAROTS by the UK, some studies of the creation of the agency state that "the glue that held ESA together was the fact that the three leading countries each got their own pet project approved" (see Harvey 2003, 163). Thus, the ESA was formed with an inherent tension, between a first-hand international cooperation and the ambition to build a transnational (rather than "denationalized") institution, and the national driving force on the second hand, pushing to contribute, which depended on the related fallout for each national contributor. Since its inception, European cooperation in space activities has been focused on the development of satellites and launchers. In April 1960, the Council of Ministers of the European Economic Community ("EEC" hereafter) recommended the creation of a space agency on a European scale, while a conference gathering scientists from eight countries of Europe was organized by Pierre Auger (researcher at CERN, the European nuclear research facility) to talk about a European organization for space research, with scientists from Belgium, the UK, France, Germany, Italy, Sweden and Switzerland. Again, discussions turned to a more formal basis around a satellite project.[4] The essential point is to note that since its beginning, the European space programme was science-led and motivated by scientist networks, as well as inextricably linked to the EEC.

If Europe's introduction to manned spaceflights dates back to Spacelab 1 (which flew with the American space shuttle in December 1983), some of its internal national programmes are older. On the occasion of a Soviet-French cooperation commission in April 1979, the very first idea of a French human spaceflight to a Soviet space station was evoked, the agreement signed in October 1979, and Jean-Loup Chrétien became the first French astronaut to effectively fly to the Salyut space station in June 1982 after training in Moscow (including familiarization with Russian space systems and learning the Russian language). CNES supervised experiments loaded for the flight, which already framed current patterns of international collaboration in the management of scientific research in manned spaceflights (see also O'Sullivan 2016).

Soviet-French collaboration continued with the *Mir* space station,[5] with Chretien's "Argatz" mission, Michel Tognini's "Antares" mission (July 1992), Jean-Pierre Haigneré's "Altair" mission (July 1993 and February 1998), Claudie André-Deshays' "Cassiopeia" (August 1996), and Leopold Eyharts' "Pegasus" (February 1998). If many of their experiments were led by CNES and on behalf of researchers from

---

4   For more details, see Harvey (2003).

5   See subsection below, titled "Structuring international collaboration through the management of experiments".

French institutes, they were also responsible for European experiments, mostly led by the German Centre of Astronautics (DLR) or by ESA. These first European flights to Mir were followed by two German flights from 1992 to 1997 (Klaus-Dietrich Flade and Reinhold Ewald) and Austrian flights (Franz Viehböck) (October 1991), which was the last space mission conducted in the Soviet period. These nation-based flights and their relative bilateral agreements turned into a European manned space programme with the Euromir missions from 1994. The programme included the Spanish Pedro Duque, the Swedish Christer Fuglesang, and the Germans Ulf Merbold and Thomas Reiter. Spacelab then appeared as the first important contribution of Europe to human spaceflights, collaborating with NASA to fund, build and perform experiments aboard the shuttle. As agreed in 1973, with a major financial contribution from Germany (followed by France), Spacelab 1 launched its first astronauts towards the end of 1983, after several delays. Experiments from Europe, the United States, and India (notably) flew aboard Spacelab missions, from plasma physics to fluid mechanics and life sciences. The Challenger disaster suspended most space programmes, including Spacelab expeditions, and finally ended in 1998 (the last European to fly was Jean-Jacques Favier in June 1996).

At that time, bases for the European organization of spaceflights were crafted through this Soviet-French collaboration, while the Soviets were one technical and logistical step ahead as regards the management of spaceflights, usability, robotic interface, and launch facilities. CNES (and, later, ESA) consequently based its procedures and management of spaceflights on Russian know-how. At the time of the French flights aboard Mir, expeditions were supported from the control centre in Korolev (near Moscow). Scientists and technicians involved in the mission consequently had to stay there for an average of two hours of manipulation at the beginning and at the end of the astronaut's flight, which imposed demands in terms of logistics. Progressively, numerical improvements allowed the installation of connection lines from Moscow to the CNES' Toulouse space centre, which allowed the transmission of data and the construction of a delocalized centre to support operations (CADMOS). At a political level, rather than at an operational one, ESA decided to implicate its member-states in term of visibility, while the only two support centres were localized in France (Toulouse, CADMOS and ATV[6] control centre) and Germany (Oberpfaffenhofen, ESA Columbus control centre), which restricted the visibility at the two main contributors. USOCs were then considered a way to increase the presence and the implication of other contributors to ESA on the international space stage, by sharing among them the organization of ISS

---

6   Automated Transfer Vehicle, an automated spacecraft performing supply missions to
    the International Space Station from 2008 to 2015.

activities. Even if these national centres have a specialization (physiology, biology, physics and so on), they serve as contact points for scientists, whatever their respective disciplines, and the latter are then oriented to the centre specializing in the relevant field. Consequently, USOCs are national contact points for the scientific community, which wants to use ISS as an experimental platform, and are also specialized in accordance with the thematic of research.[7]

While ESA endowed itself with operation support centres, representative of its national multiplicity and its geographical vastness, it was looking for a relatively autonomous programme in which its contributions would not be restricted to a status of being a temporary host. In January 1985, ESA decided to commit to the funding and the construction of a European module—Columbus—which would be a research facility for (mainly, but not exclusively) European experiments. Meanwhile, Russia and the United States were working on a new space station, NASA having expressed a desire for an international project to avoid competition since the early 1980s. A first memorandum was signed in 1985 among member-states, with a major contribution by NASA (71.4 %), followed by ESA (12.8 %) and JAXA (12.8 %), and by CSA (3 %). Negotiations then underwent complications, from American Congress cuts in the station's budget to a bilateral agreement among NASA and ISA (the Italian space agency) outside the ESA jurisdiction, with the construction of modules by Italian industries (Alenia), which would make of Italy one of the three main contributors to ISS (with Germany and France) and would reserve special seats at the ISS. Nevertheless, in the first few versions of the agreement, Europe, Canada and Japan still played a very restricted part in the ISS (sometimes also called "Alpha"), and the first international meeting gathering all member-states took place in November 1993, after the United States and Russia had agreed to the design of the station. It was only two years later that negotiations with regard to a European contribution to the station became serious, with a participation of 10 %, mainly assured by Germany (41 %), France (28 %), Italy (19 %). The UK did not participate, and the other ESA member-states contributed up to 3 %. It was then agreed that Europe's contribution would take the form of a research facility (Columbus), a data management system, a robotic arm, and the Automated Transfer Vehicle (ATV).

The project of the ISS was originally planned for a launch scheduled in 1992 (the European laboratory Columbus would then have been loaded for celebrating the pentacentenary of Christopher Columbus's expedition to America). Delays occurred partly because of technical and budgetary issues. The first ISS module—the Russian

---

7   This paragraph is the result of an interview with a director at CNES, CNES 1, 21.07.2016.

Zarya ("Dawn") module—was launched in low Earth orbit in November 1998, and its manned occupation started in October 2000 with Expedition 1.[8]

4.2.2   Structuring International Collaboration through the Management of Experiments

While the training of astronauts and cosmonauts called to fly to the ISS is divided among Russia, the United States, Germany and Japan, depending on the infra-structures studied, the management of activities once the astronaut is on board is internationally coordinated as well. Because the space station comprises two main segments, one led by the Russian space agency and the other—hosting European, Canadian, American and Japanese facilities—led by NASA, all European activity aboard the station is ultimately led from the Marshall Spaceflight Centre (MSC) in Houston. European activities are, therefore, restricted to experiments conducted in the European facility: The Columbus laboratory, launched in 2008. Regulations of activities aboard Columbus impose the stipulation that any experiment performed within it has to be managed from the ESA control centre in Oberpfaffenhofen (near Munich, Germany), even if the astronaut performing the activity does not belong to ESA. The experiment PK4 is quite representative and highlights the delegation's raison d'être. Experiments in plasma physics resulting from a cooperation among German[9] and Russian[10] scientists, PK4 (Plasma Krystall-4) and supported by the German Centre for Astronautics and Aeronautics (*Deutsches Zentrum für Luft- und Raumfahrt*, DLR), aims to understand plasma (the fourth state of matter after solid, liquid and gas) crystal formation and behaviour without gravity, focusing—for example—on ion drag force measurements, measurement of particle charge, or on self-organization and non-equilibrium phase transitions of plasmas. After several parabolic flights from August 2002 to December 2004, the experiment was finally launched aboard ISS in March 2015, supported by the CADMOS payload, inside the European Columbus laboratory. Indeed, the experiment was exceptionally operationalized from the Toulousian operation centre—the latter specializes in physiology and the support of medical experiments—because of the need to use its payload. This experiment has two specificities: It is a European experiment conducted with a Russian cosmonaut (which means that the experiment is con-

---

8   The crew was composed of the American William Shepherd and the Russians Serguei Krikalev (currently head of human spaceflights at Roscosmos) and Yuri Gidzenko.

9   From Max Plank Institute.

10   From the Institute of High Energy Densities (IHED) of the Russian Academy of Sciences.

ducted on Russian crew time, each agency having a defined time of activity for its crew per increment[11]), and its operational support is not led from the main ESA control centre (in Oberpfaffenhofen, Germany) via the Eurocom operator team (cf. below). During two weeks per year, activities dedicated to the experiment are spread over several steps in accordance with the other scheduled activities on crew timelines. To take the example of one of its campaigns in June 2016, the week began on a Sunday by starting the instrumental device (depressurizing, creating a void inside the caisson where plasma fusion would occur during the experimental phases, and so on). There would not have been any crew activity before Tuesday; the Russian cosmonaut would start the loaded laptop in Columbus and launch the software. Wednesday and Thursday were marked by more intense activities, since they were dedicated to the formation of crystals through plasma fusion, with the cosmonaut aboard.

Three types of international coordination of activity within Columbus can occur (mainly bilateral). The first one is the case detailed above, illustrated by PK4: The experimental (i. e. the scientific) team is European, as are the operators working on the operationalization of the scientific project from an ESA centre (i. e. developing the procedures to be used by the crew). One should specify that PK4 is an exception regarding several points, including the fact that the communication between the ground and the station avoids the regular procedure: partly because the activity runs with a Roscosmos cosmonaut rather than with a European astronaut, the Eurocom team of operators (usually in charge of the communication with the station for ESA activities) let the USOC operator (who worked with scientists and industrials on the operationnalisation of the project) guide the cosmonaut through the procedure, shadowed by representatives of the investigators team. ,allowed to talk to the crew at ESA). ,the This exemption to the regular procedure remains exceptional, and the usual way is as following: either the crew *and* the experiment are european, then the activity will be performed in Columbus (the ESA's loaded laboratory) and the "Eurocom" team of operators (located at the main European control center in Oberpfaffenhofen) will lead all the communications between the ground and the station , with the required bureaucratic organization and its resulting decentralization; or the European crew is charged to perform an experiment led by another space agency in its related research facility (NASA's Destiny, JAXA's Kibo or Roscosmos laboratories [*Poisk* or *Ravssek*]), and the activity will

---

11 An increment is a six-month period structuring the ISS planning. An expedition or astronaut/cosmonaut mission is usually extended between two increments. For example, the "Proxima" mission of the ESA astronaut Thomas Pesquet (aboard ISS from November 2016 to May 2017) was extended on Increments 50 and 51.

then be sustained by the nation-related control centre (for instance, an American or a Canadian experiment would be conducted at NASA facilities). This second kind is generally an opportunity for space agencies that do not have frequent flights, such as the Japanese space agency. Canadian astronauts are, for their part, always attached to the American space agency (NASA) from the very first part of their training. A third kind of arrangement occurs when a European astronaut performs a European experiment in Columbus, the European facility. The latter is the most frequent kind of management of experiments aboard the station, which supposes a concern for national handling (multinational, in the case of ESA). In very rare circumstances, a fourth type is observable, where neither the astronaut or the experiment are European, but the activity is conducted within Columbus because it requires technical or equipment infrastructures not available in other station facilities.

Only scientific experiments and educational programmes are led at the national level, at the opposite end of maintenance activity or extravehicular activities (EVA), i. e., activities outside the station performed in tandem to collect data on external instruments or to provide repairs, on an average during six hours, without non-nominal events. The latter are—for their part—managed from the Korolev (Tsup) or Houston space centres, whatever be the nationality of the astronaut (as long as it is not a Russian EVA, which would be handled by Roscosmos).

4.2.3   Framing International Collaboration: A Typology

Hence, strategies for maintaining a European political order is contradicted, first of all, by the inherent internationality of current space programmes—not only by budgetary issues, but also because international collaboration turned to be the fundamental concern of space programmes.[12] When one goes beyond 350 km above the Earth for a six-month stay, and the conditions of flight are quite perilous and mandate important teamwork, matters of geopolitical discord should "stay below".[13] Before that, they stayed together for a year aboard the station as a part of a special expedition by NASA and Roscosmos, when the Russian cosmonaut Mikhaïl Kornienko and the American astronaut Scott Kelly rushed to deny that the increasing enmity between their governments could threaten their project of international

---

12   Interview with a former ESA director-general, 12.01.2017.

13   Andreas Mogensen, interview of Expedition 45 "Visiting crew conducts news conference in Russia", 11.08.2015., NASA website: https://www.nasa.gov/press-release/media-accreditation-open-for-space-station-crew-news-conference-interviews.

cooperation. The latter is, then, an aim as well as a necessary condition for the proper functioning of the station. One could identify three main types of international collaboration in the management of European experiments aboard the ISS.

1. *Logistical cooperation*, wherein the collaboration is motivated by the station infrastructure, just as it would have been the case during the ESA "Short Duration Mission" (SDM) in September 2015 among European and Russian space agencies. Soyuz spaceships cannot stay docked at the station longer than six months (i. e. the common duration of missions aboard the station) after having conveyed a crew, partly because of the composition of motors. Consequently, ESA and Roscosmos planned the SDM: while Scott Kelly (NASA) and Mikhaïl Kornienko were on board for a unusually long-duration flight of one year, it was necessary to switch spaceships in order to avoid technical issues on the Soyuz. The SDM was then the opportunity to switch spaceships to mitigate those technical risks resulting from this "Year in space" mission: if Scott Kelly and Mikhaïl Kornienko were not supposed to fly back until twelve months, two other crew members had to make the Soyuz's rotation work, hence the ten days mission for which the ESA Danish astronaut Andreas Mogensen and the Kazakh cosmonaut Aydin Ambetov launched with a Soyuz intended to replace the one that Scott Kelly and Mikhaïl Kornienko used to fly to the station.

2. *Instrumental cooperation* refers to cooperation motivated by the utilization of materials and instrumental devices. In the case of the robotic experiment METERON,[14] relations among B.USOC (an ESA centre located in Belgium) and the Marshall spaceflight centre were dictated by the utilization of an American communication network requested for the experiment—consisting of commanding a rover located at the ESA's European Space Research and Technology Centre (ESTEC) in the Netherlands from the station—in order to improve the telecommunication systems for the exploration of Mars.

3. *Structural cooperation* is, for its part, related to the organization of space activities. At the end of his mission, Andreas Mogensen was charged with assembling a device as part of a physiology experiment (MARES[15]) led by the Toulousian ESA's centre CADMOS.[16] He was assisted in his task by the Russian cosmonaut Sergueï Volkov for several hours. This cooperation occurred because of the training: Volkov was the only crew member to have undergone training

---

14 Multipurpose End-To-End Robotic Operation Network.
15 Muscle Atrophy Research and Exercise System.
16 Centre d'Aide au Développement des activités en Micropesanteur et des Opérations Spatiales.

for this operation. As it is difficult to train every astronaut or cosmonaut for every operation,[17] the way in which the training is organized orients the type of cooperation. As a result, the modalities of cooperation implant themselves, often with reference to the composition of the crew for the flight to the station: Astronauts and cosmonauts who share a spaceship also share the training. This last type of cooperation also refers to what insiders call "cooperation potential". The term is used to evoke operations that gather scientists from different countries, and even from different space agencies, on a common research project. Cases where no cooperation potential is obvious are mostly situations where a research is already in progress as a part of another space agency's funding on the same research issue. The space agency then creates a network with scientists working on similar questions in order to reduce fees.[18]

4.3   Organizational Conditions: Manufactoring Conditions of Political Power

From an organizational standpoint, political power requires a material base. Indeed, European political cohesion requests a stable state political power that would be institutionally and axiologically defined. Consequently, bureaucratic organization of state activities (used as a model within each governmental agency, including space agencies in their management of scientific experiments as well as all activities aboard the ISS), should be taken as a central issue to understand the modalities of construction and the exercise of this political power. After a detour involving Antonio Gramsci's theorization of political crisis leading to a rigidification of the state apparatus, I will present the adaptation that Cédric Durand and Razmig Keucheyan propose to explain the UE's governance crisis in the aftermath of the 2009 financial crisis—turning Gramsci's caesarism exercised by the repressive state apparatus into a bureaucratic form of caesarism as maintained by an international institution—and adapt it to the European space agency to explain the function of proceduralizing experimental protocols in state intervention within the scientific process (turning original financial hegemony of "bureaucratic caesarism" to a scientific hegemony, ultimately contributing to state political power).

---

17   Interview with the team leader of CADMOS, 05.11.2015.
18   Interview with ESA 1, 20.10.2015.

4.3.1   Towards a Bureaucratic Caesarism

According to Gramsci, the caesarist solution to a crisis highlights the fact that every democratic state is constituted by institutions that are more or less democratic, and in situations of crisis, the more democratic institutions step back in favour of the less democratic in order to palliate the lack of governance and the caesura among political state and civil society consequently provoked. Caesarism then alludes to a rigidification of the State's apparatus and of its action, through the adoption of an authoritarian solution to a crisis[19] (as said above, through the repressive State apparatus: Its army). It is, then, the "tendency of democratic regimes to manifest authoritarian leanings in time of crisis" (Durand and Keucheyan 2013, 75). The "bureaucratic caesarism" of the European Central Bank—elaborated by Cédric Durand and Razmig Keucheyan—highlights a case of caesarism that is no longer exercised by the State through the coercive action of its army, but by an international institution, thanks to its bureaucratic organization. Although fragmented, Europe remains no less a political entity, the unity of which is guaranteed by the bureaucracy which, through one of these organizations (such as the ECB), is able to apply an authoritarian Caesar function on the financial products of the European Union. Thus, two elements circumscribe the phenomenon of bureaucratic caesarism. First, a crisis from which it can develop, and secondly, a bureaucratic organization allowing an institution to exercise a hegemonic political power driven by the centralization of decision-making.

Because of the economic crisis muted in an "organic crisis", democratic institutions go back upon the governance stage, and an authoritarian governance emerges in order to maintain a political order. The financial hegemony refers to the exercise of a monopoly on finance, maintained—in the case of the bureaucratic caesarism of Durand and Keucheyan—by the European Central Bank. Since the monetary crisis of 2009 destabilized the economic sphere as well as spread to the political sphere, the regulation of finance has to be assured by an international organization.

Bureaucracy is specific to the exercise of political power in the case of the capitalist State. For this reason, bureaucracy governs production relations constituting

---

19   To Gramsci, the crisis that could lead to a form of caesarism is called "organic crisis". It occurs when a disequilibrium in the economic sphere infects the political sphere and upsets the social order previously established by the "historical bloc" (e. g. a symbiosis of economic, cultural and political structures at a particular time of the capitalist production, which acts like a fortification protecting the political power of the state and of the civil society, and generally avoids such a contagion). See Gramsci (1978).

the "institutional materiality of the State", in Poulantzas's words (see 1972, 155).[20] A bureaucratic organization is therefore (notably) defined in accordance with the delimitation of competency fields, a ranking of individual functions coupled with a rationalization of careers, a codified (and codifying) regulation of the professional life (as well as the private life under certain circumstances), as well as rationalized tools of labour (see, for example, Bourdieu 1971; Wilson 1989). As part of the management of professional activity, procedures maintain the institutional base of the organization throughout the reproduction of its practices, and certify the rolling of its ranked positions and fragmented competencies. Because it consequently structures internal constraints at the practice, bureaucracy permits what Noam Chomsky would qualify as "manufacturing consent", necessary for any monopoly in the exercise of a legitimate symbolic power. Thus, the specificity of the scientific hegemony produced by the ESA (and other space agencies) is based upon a division of production labour fluctuating between the monopoly over the management of scientific research aboard the ISS, and that of the very definition of scientific knowledge. States participating in ISS activities can, therefore, constitute a political entity beyond national boundaries because of their common bureaucratic organization certifying their coordination, but also the building of an internationalized monopoly in the regulation of scientific experimentation.

Since they manage experiments in the contexts of manned spaceflights and their activity norms—long temporalities, safety rules, ranking of positions and functions, formalization of protocols, construction of specific instrumental facilities and so on—space agencies participate in the value accorded to experiment results. The scientific hegemony specific to space agencies is then related to the development of a monopoly detained by an international organization over the experimental practice. Even if this caesarism is inherently international, the case of ESA illustrates the role of caesarism in the European construction: In the ambition to build a "European identity", astronauts are ambassadors of a "unified Europe", gathered around its internal differences through its know-how and its scientific production made in Europe.[21] Compared with the financial caesarism of the E.U. evoked above, procedures and flight rules are analogous to memorandums. The bureaucratic caesarism of the ESA consequently refers to a tendency to centralize most of the decisions regarding scientific policy, its management and its practices.

---

20 For Poulantzas, the state is capitalist if it is an instrument of the reproduction of the capitalist system of production, coordinating interests of "fractions of capital" (to be understood as specialized sectors of production within the society: financial capital, military capital, industrial capital, and so on). See Poulantzas 2000, second part.

21 Interview: ESA 13, former ESA astronaut, Paris, 09.11.2015.

More than a "relative autonomy" of the State resulting from the division of expert labour and consequent struggles among actors, the bureaucratic caesarism highlights[22] production relations where State and science together circumscribe what is generally the core of science (i. e. the experimental practice).

4.3.2 Delegation and the Scientific Hegemony: Intervening in the Scientific Field

As explained in a previous section, financial hegemony refers to the exercise of a monopoly on finance, maintained—in the case of the bureaucratic caesarism of Durand and Keucheyan—by the European Central Bank. Concerning human space flights and scientific research aboard the space station, institutions from political and scientific fields struggle for the monopoly over the experimental work. The latter is central in the scientific process because it is that step of scientific production during which the confrontation with the empirical data will provide the credit and the legitimization to scientific discourses and knowledge. This is what sociologists and historians of science often refer to as "administration of proof", which is inherent in the institutionalization of science (see, for example, Galison 1987; Shapin and Schaffer 1985). This is also related to the constitution of laboratories as privileged places producing legitimate knowledge.

Scientific research aboard spaceflights is based upon two kinds of delegation, which form the conditions for the scientific hegemony specific to the ESA's bureaucratic caesarism. First of all, nation-states delegate their political power to an international organization (in the case of the European space agency) or to one of its internal agency. The space agency consequently exercises a political regency through the bureaucratic organization of its activities. On the other hand, scientists delegate the experimental part of the scientific process to operators (operationalizing the experiment) and astronauts (performing the experiment aboard the station). Such delegation results from the fact that research aboard space stations is managed by a space agency from the first steps of the projects to the conditions of the experimentation, but also because the experiment is not a part of the scientist's work anymore, but the core of the astronaut's activity during his/her mission (including around 80 experiments per flight lasting six months). After a process of selection, experiments go through several steps of "spatialization", during which scientists and space agency operators collaborate to adapt scientific ambitions to operational

---

22 Framed by Louis Althusser, the idea of a "relative autonomy" of the state is also discussed in detail in Nicos Poulantzas's writings.

possibilities, depending on the loaded infrastructures and safety conditions in the station. The experimental protocol is then translated into procedures confectioned by operators, so that the astronaut will repeat the protocols during his/her training. This delegation is then the logistical condition of a political concern: Legitimizing the human occupation of circumterrestrial space (Bell and Parker 2009).

Because outer space is quite unsafe (cosmic rays, microgravity producing muscular atrophy, decalcification, and so on), scientists cannot perform their experiments themselves and have to delegate these to trained professional astronauts. Thus, because space agencies manage experiments under microgravity aboard the ISS and because their norms (long temporalities, safety rules, the hierarchical structure of the bureaucracy) intervene in the way science will be defined (as to what constitutes scientific progress, what is a valuable scientific project and how it should be realized), it is possible to understand here how a state intercedes in the experimental process, which attributes credit to scientific theories. Nevertheless, conditions for this intervention (which represents a specific kind of relationship among state and science) also depends of the capacity of internal cohesion of a state through one of its agency, and to the materiality the latter imparts to its political power.

Nevertheless, the proceduralization of experiments in this manner is far from being a novelty in the scientific process, as one should know. Such bureaucratization (for which procedures constitute the most obvious objectivization and the main tool of operationalization) results from two phenomena: Neoliberalization and the evolution of the division of labour.

4.3.3   ESA as a "Coordinating Agency" within the European Territory

To pretend to its exercise, political power requires a relatively stable state that would be defined institutionally and axiologically (see, for example, Wilson 1989). Consequently, the bureaucratic organization of state activities (used as a model within each governmental agency, including space agencies in their management of scientific experiments as well as all activities aboard the ISS) should be taken as a central issue to understand the modalities of construction and the exercise of this political power.

The concrete conditions of political power as regards the organization of activities within a state quite often lack an analytical aspect in the literature, with most of the latter's theoreticians often preferring a microanalysis, neglecting the concrete modalities of political power (Arrighi 1983). Among some exceptions, the Indian Marxist sociologist Vivek Chibber proposed a postcolonial apprehension of the bureaucracy in the process of the construction of the state (2002; 2003), emphasizing the bureaucratic rationality and the latter's function as a condition for the construction of an effective political power. Following the thesis of Peter Evans—who explained how the bureaucratic system provides social and organizational cohesion that participates in a stable State power (in the sense that it standardizes individual values and professional practices)—Chibber then posits that bureaucracy is not sufficient to produce internal cohesion in itself, so that political power could be exercised. Internal cohesion requires what Chibber introduces as a "nodal agency", a coordinating agency linking all the other governmental agencies composing the state, in order to impose priorities, homogenize their differences of interest and, in the end, regulate the relations of authority among them. The coordinating agency consequently "has the function of resolving intrastate conflicts" (ibid., 958.). Indeed, "conflicts arise because the state is a complex amalgamation of agencies, charged with distinct functions, having domains that are frequently overlapping, and often compelled to compete for resources. Hence, interagency conflicts will arise because of following the rule that govern their reproduction, not because of a departure from the rule" (ibid., 957–958).

A space agency is precisely in such a position, such that it can assume this function of a "coordinating" agency within a state: Coordinating distinct social sectors (science, industries, economy, politics) and their relative institutions. Managing (and funding) experiments aboard the ISS require replacing or collaborating with organizations generally funding scientific research (National Science Foundation, Centre National de la Recherche Scientifique and so on), proceeding to the selection of projects across countries or international networks after diffusing a call for projects: for instance, the European Science Foundation (ESF hereafter), which is also charged with proceeding to a double-blinded peer-review on the account of the ESA for the evaluation of scientific projects, or creating research partnerships through the scientific community and among scientific and industrial organizations. Unlike other agencies of research funding, space agencies such as the ESA intervene directly in the scientific process (as mentioned above), by imposing a necessary frame orienting the operationalization of protocols, which corresponds to the safety requisites of the space station. Consequently, the space agency must regulate a division of labour that does not occur outside the field of sciences in microgravity. This specific division among institutionalized sectors is then structured through the bureaucratization, which homogenizes different kinds of expert knowledge and practices, and coordinates what ordinarily belongs to the jurisdiction of disparate governmental agencies. Such coordination of the scientific process led by the state (through the space agency) and, in the meantime, empowering other UE institutions (such as the ESF) allows it to build a material base for a political power to exercise in the interstate system as well as contributes to its internal cohesion.

4.4   Axiological Foundations: The Symbolic Side of the Moon

To Peter Dickens, if there is a political competition among states for resources in outer space, the latter is led without the repressive state apparatus (the army) partly because there is no state property to fight over in outer space. Official agreements from the United Nations stipulate this interdiction of property on celestial bodies (such as the Moon). However, the blind spot in the work of Dickens is also the missing point of Marxist theories of political power: The symbolic conditions for its materiality, governed by the way science has been defined through a historical and social process, drive to its a priori depoliticization. This section is focused on the mechanisms through which the European contribution to the ISS and science aboard the station lead to the constitution and the evolution of a European political power (as well as of political power specific to the ESA's member-states), as well as how it results from a Western epistemology and a social legitimacy of the scientific production leading to its definition as depoliticised.

4.4.1   Making Science as Depoliticized

Scientific research in manned spaceflights contexts has been introduced in the space activities landscape since the very beginning of the "humanization of space", with the first space station permitting a continuous occupation of space: The Soviet *Mir* station ("Peace", 1986–2001). Mir resulted from the first partnerships between the Soviet Union and the United States since the Cold War's stalemate, with France in the middle. While during the Cold War, the stalemate of the competition between the two superpowers to justify the conquest of space ended up raising the scientific practice as the major concern of civilian space programmes—because "[p]rotecting the civil society and the freedom of citizens is deemed a better alternative to 'macho ads with missiles and fighter planes'" (Slijper, cited in Dickens 2009). This process implied a prerequisite: that the political issue of space programmes may be subverted by the "cultural authority" of science. As mentioned above, manned spaceflights participated in the resolution of a crisis of capitalism, reframing the territory of its relations of production. Be that as it may, this process requires the acknowledgment of a public utility, which could be found in the injunction to put an end to the military justification of spaceflights in the early post-Cold War period. Indeed, regarding the space environment (including microgravity and solar rays), and the impossibility of reproduce its conditions on Earth for long-duration experiments, the relevance of developing science in space may seem obvious. One could enumerate three determining factors, thanks to which the scientific practice became a main concern in the legitimacy of manned spaceflights, for taxpayers as well as for neighbouring states. First, science forces international partnerships, since the latter is necessary for a "big science", demanding technical and financial resources (1963). Science is also inherently international since it includes "more sophisticated relationships, such as joint research programs involving institutions and individuals across countries, and internationally shared research facilities, such as giant optical telescopes, space launching facilities, and macroaccelerators in the field of particle physics" (Ancarani 1995, 643–54). Moreover, one should note that two of the main characteristics of this "big science" are the internationalization of social networks among the scientific community and an expansion of the instruments through space. Secondly, science is invariantly defined as detached from any political interest; it is, consequently, the counterpart of politics in a dualistic opposition. Indeed, interviews conducted as a part of the empirical inquiry show that political actors and operators from every space agency involved in the international space station[23] mark a distinction between the first spaceflights and their political concerns related to intergovernmental conflicts on the one hand, and current or future spaceflights concerned by the exploration, scientific and technological development and international cooperation on the other.

By so defining science as depoliticized from the outset, the fieldwork takes a look at the conditions of what Gordon Gauchat (2001) and Paul Starr (1982) define as the "cultural" or the "cognitive authority" of science: If science is the opposite of conflict, it is then because it relies on a rigorous empirical method based on a confrontation of theories and hypothesis to the reality through a process restricting subjective bias. Nevertheless, science is also structured by a specific ethos or normative corpus, in which the objectivity and its related axiological neutrality predominate. From that, science owns its "authority" to discourses and knowledge, in the sense that its discourses and its practices would not be questioned and would be legitimized on an a priori basis. Processes validating knowledge are central in feminist and postcolonial critiques of science and technology since it is there that discourses and experiential knowledge of oppressed people are discriminated, as

---

23 American, Canadian, European, French and Russian space agencies, from September 2015 to July 2016.

the case of the Black feminist epistemology, described in detail by Patricia Hill Collins, could be delineated (2000).[24]

Hence, scientific knowledge would be a mainstream knowledge (maintained and produced by a dominant standpoint with regard to the class, the sex, or the race) rather than the universal voice, while the neutrality of objectivity remains a major issue in various theoretical streams resulting from Marxism and poststructuralist critical theories that emerged in the late 1970s. For it seems plausible that this a priori legitimacy of science could be a useful tool to forge a pacified state practice serving the common good. What would have been more logical than making the main concern of science the spaceflights which were seeking a non-political reason that would have forced an international cooperation impossible to obtain on Earth? Hence, the project of the international space station emerged in the early 1990s: To craft peaceful international affairs, to follow the rules imposed by the United Nations (UN 2002; 1979), and later, to launch a global economy through the development of an out-of-the-Earth market.

4.4.2   A Western Epistemology

This process of depolitization iscan be seen as dependent from the historiography of science as it was developed in modern Europe, i. e. the way scientific process has been conceptualized and the role attributed to science and technology in political mechanisms of modern societies. In the case of the Soviet union, a stream in history of science—often led by Soviet party's leaders such as Nikolai Bukharin—highlighted how science, as a provider of ideological legitimacy, could have been conceived as a tool in the stability of the post-revolutionnary society (cf. Omodeo 2016; Bukharin 1934). Crafting an epistemology and a history of science based on the historical materialism of Karl Marx, the notion of "objectivity" or of "truth" in the apprehension of the natural reality are questioned: according to those theories, the objectivity is not a matter of axiological neutrality but is related to the adequation with the Marxian subalterns' epistemic privilege. That means that the "truth" does not refers to a "point of view of nowhere" (Nagel 1986) but to the one of the dominated in a system of power social relations (according to the principale that the subaltern will doesn't have any advantage to contribute to the reproduction of the system of domination as it is established).

---

24   A social history of critical epistemologies would deserve an entire article, which is not the subject of the present one.

The epistemology as it has been developed in Western Europe is, as one should know, quite the opposite. The historiography of science and theorisation of the scientific process (for example, regarding paradigm shifts and scientific revolutions) as it is thought by "classic" references such as Alexandre Koyré, Thomas S. Kuhn or Karl Popper in the 20th century is characterized by the axiological neutrality, an objectivation processs leading to the deletion of any trace of subjectivity—what would have been qualified of "ideologically correct" epistemology (see for example Pollock 2006). If the reason of developing such an epistemology supposed to support the socialist project in Soviet union is partly explained by the diffusion of Marxism and need for an ideological base for rebuilding a socil order after the soviet revolution, the reasons of the Western epistemology is mostly explained by the colonial past of Europe (and United States) and its pretentions to/of universality.

Science would have contributed to the constitution of a "cultural hegemony", the criticism of the bourgeois science is part of struggles among ideologies while the Western epistemology would be the anticommunist reaction to the Soviet historiography of science (Omodeo, 2016). Without concurring with this view, it remains interesting to look at how political allegiance or opposition are a part of the historical construction of a definition of science where the latter berefits of any political substance. All the interest of the analysis is here to look at science as an ideological foundation of the European political community and its social and political order, just like science in the emerging Soviet union.

4.4.3   Science in the Capitalist Relations of Production

In the case of human spaceflights in the post-Cold War era, the question arises as to what means could there be under capitalism to overcome the accumulation of capital and the constitution of power relations among States for finance capital. Since scientific production is the major concern and the fundamental purpose of space stations, and the aim is the permanent occupation of outer space, it may seem obvious—or at least relevant—to find out the functions of science in the political power process, which are dependent on the (uneven) development of capitalist production relations, as described above, with the specificity that it cannot occur through military action. Science, therefore, appears as a central concern to understand political competition in nonmilitary contexts. Already in texts by the thinker of the cultural hegemony, domination is sustained by a dichotomy opposing coercion and consent—furthermore in societies that distinguish between State and civil society. These distinctions would be a "historical necessity related to the development of capitalism, from a recodification of warrior aggressiveness to concurrent aggressiveness" (Bentouhami 2014, 101, my translation from the French). Consequently, this supposes a rethink of the notion of authoritarianism through the transformation of political power, increasingly exercised by economic competition.

The development of spaceflights is inherent to the emergence and resolution of one of the crisis of capitalism, thanks to what the Marxist geographer David Harvey names a "spatial fix".[25] Largely inspired by Henri Lefebvre, but also critical about his work focused on capital's production of space, Harvey reckons capitalism does not resolve any of its crises, but rather shifts them through space or time. Indeed, one of the permanent features of capitalism is that "if the surplus cannot be somehow absorbed then they will be devalued" (2006). This kind of crisis generated by overaccumulation is inevitable, and consists of systemic stages. Most of the time, These crises are relieved by what Harvey describes as "temporal shift" or "spatial fix": the first is the uptake of the production of surpluses in long-term projects so that investments earnings would be spaced out over time, as the second one is reached by displacing (exporting or dispersing) these surpluses through space into new or renewing territories of the exploitation of resources. Thus, following Marx, Harvey explains how the overaccumulation of capital emerging in the 1970s led to the adoption of neoliberal public policies (2003: 184) and requested the territorial renewal of the circuits of circulation for capitals (including new workforces, new materials of production and to product, a new market, and so on), in lands that have been ignored by capitalist relations so far, so as to spread or rejuvenate the spaces of capital.

The institutionalization of spaceflights (including manned spaceflights) precisely consists of such an expansion through space. In such a way that Peter Dickens and James Ormrod highlight (following Harvey) what they call the "outer spatial fix" (2007), meaning that the "humanization of the cosmos" allowed the reproduction of the capitalistic circuits of production. This reproduction (and the renewal of its workforces, materials and market) proceeds by establishing new industrial partnerships, the creation of specialized and competitive companies, and the exploitation of natural resources (such as space mining). While the crisis of capitalism that occurred in the 1970s (characterized by a permanent deceleration of growth rates and an increasing financial hyperplasia) marks a "redeployment of capitalism", in Samir Amin's words (2001), David Harvey notices that it is also the starting point for a new kind of state power: An "imperialist" dynamic of capitalism distinguished by a process of accumulation by dispossession. This constitutes, for Harvey, a response

---

25 For Harvey, the spatial fix intervenes in the resolution of crisis of capitalism because capital is territorialized and because the resolution of crises goes by investing in new spaces of production or regenerating old relations of production). See Harvey 1982.

to the decreasing profitability and risks of overaccumulation from the early 1970s to the present time. In so doing, the geographer aims to understand the intimate links among the neoliberal imperialism and the expansionist military policies maintained by the second Bush administration. As long as it legitimizes manned spaceflights, science aboard space stations aims to stimulate the collaboration and to benefit capitalism, so the expansion of its territories of production can evade one of its systemic crisis (caused by accumulation and/or exceeded costs in the production of value) or, at least, resolve it through a spatial fix.

4.5   Conclusion

This chapter focused on mechanisms though which science provided legitimacy for manned space activities since their institutionalization: First, by being a part of the capitalist process of production, and second, by being used in peaceful policies and justifying the human spaceflight plans—in order to launch and maintain European space policies. It also aimed to present the necessity to analyse the anatomy of political power as well as its prerequisites, through a concrete way to organize state activities and constitute a political power, as well as through a symbolic condition resulting from the cultural authority of science (which remains the main concern of spaceflights).

As described in detail above, European integration is largely based upon the model of a Europe of science, technology and research, which is sustained by the European space programme (in our case, in manned spaceflights), thanks to organizational conditions led by the bureaucratic management of experiments aboard the space station, and axiological foundations related to mainstream definitions of science (as depoliticized from the outset) and its place within capitalist production relations structuring the state in Marxist theories. Indeed, in the case of the European space agency and its management of scientific experiments aboard the international space station, the State meddling within the scientific process under a caesarist shape results from four features: the constitution of a scientific hegemony, a logic of delegation, the proceduralization of protocols (administrating the credibility to scientific production) and the resolution of an overaccumulation crisis through the expansion of production territories into outer space. The crisis from which this caesarism results is then intimately related to the inherent contradictions of capitalism (or, in a more precise way, to its systemic crises).

Because it constitutes the most advanced stage of capitalism (considering the expansion of the capitalist global spaces off the Earth), and because it presents an imbrication of science, politics and the economy which cannot be observed elsewhere in such an outspoken manner, there is no doubt that space activities put a relevant spotlight upon European space policies and the evolution of the relations of production structuring capitalist processes, and contribute to an analysis of the political power of states in a globalized area of activities. Moreover, the permanent evolution of space-related activities, especially with regard to human spaceflights and perspectives of (commercial) space travels through the solar system, not only promises valuable improvements for the study of capitalism, but should also allow reworking conditions of political power anchored in the current production relations, the phase of (financial) capitalism, and the power relations on the international stage.

Since manned spaceflights act out structural relations among State, science and market that are particularly explicit within the European space of production, they seem to constitute a laboratory for a dialectic among the economy and politics, without forgetting the consideration of science as a cornerstone of these relations. In such periods of demilitarization of the coercive action of state and the transformation of the forms of political violence, there is no doubt that one should take the function of science as a part of the process of state-building, in the sense that it structures its symbolic base, and is a full-fledged part of the relations of production structuring the "institutional materiality of the State".

References

Amin, Samir. 2001. *Au-delà du capitalisme sénile: Pour un XXIe siècle non Américain*, Paris: Presses Universitaires de France.
Ancarani, Vittorio. 1995. Globalizing the world. In *Handbook of science and technology studies*, ed. Sheila Jasanoff et al. Thousand Oaks, CA: Sage Publications.
Arrighi, Giovanni. 1983. *The geometry of imperialism: The limits of Hobson's paradigm*. New York: Verso.
Bell, David, and Martin G. Parker. 2009. *Space travel and culture: From Apollo to space tourism*. Hoboken: Wiley-Blackwell.
Bentouhami, Hourya. 2014. De Gramsci à Fanon: un marxisme décentré. *Actuel Marx* 55: 99–118.
Bourdieu, Pierre. 1971. Genèse et structure du champ religieux. *Revue française de sociologie* 12: 295–334.
Brice, Laurent. 2013. "Du laboratoire scientifique à l'ordre constitutionnel." Analyser la représentation à la suite des études sociales des sciences. *Raisons politiques* 50: 137–155. doi:10.3917/rai.050.0137.
Bukharin, Nikolai. 1934. *Historical materialism: A system of sociology*. New York: International Publishers.

Chibber, Vivek. 2002. Bureaucratic rationality and the developmental state. *American Journal of Sociology* 107 (4): 951–989.

Chibber, Vivek. 2003. *Locked in place: State-building and late industrialization in India*. Princeton: Princeton Univ. Press.

Collins, Patricia, H. 2000. *Black feminist thought: Knowledge, consciousness, and the politics of empowerment*. London: Routledge.

de Solla Price, Derek. 1963. *Little science, big science*. New York: Columbia Univ. Press.

Dickens, Peter. 2009. The cosmos as capitalism's outside. *Sociological Review* 57: 66–82.

Dickens, Peter, and Ormrod, James. 2007. *Cosmic society: Toward a sociology of the universe*. New York: Routledge.

Durand, Cédric, and Razmig Keucheyan. 2013. Un césarisme bureaucratique: une lecture gramscienne de la crise Européenne. In *En finir avec l'Europe*, ed. Cédric Durand. Paris: La fabrique.

Durand, Cédric, and Razmig Keucheyan. 2015. Financial hegemony and the unachieved European state. *Competition and change* 19 (2), 129–144.

Elzinga, Aant, and Andrew Jamison. 1995. Changing policies agendas in science and technology. In *Handbook of science and technology studies*, ed. Sheila Jasanoff et al. Thousand Oaks, CA: Sage Publications.

Galison, Peter. 1987. *How experiments end*. Chicago: Univ. of Chicago Press.

Gauchat, Gordon. 2011. The cultural authority of science: Public trust and acceptance of organized science. *Public understanding of science* 20 (6): 751–770.

Gramsci, Antonio. 1978–1992. *Cahiers de prison*. 5 vol. Paris: Gallimard.

Granjou, Céline. 2003. L'expertise scientifique à destination politique. *Cahiers internationaux de sociologie* 114: 175–183. doi:10.3917/cis.114.0175.

Harvey, Brian. 2003. *Europe's space programme. To Ariane and beyond*. London: Springer.

Harvey, David. 1982. *The limits to capital*. Oxford: Basil Blackwell.

Harvey, David. 2003. *The new imperialism*. Oxford: Oxford Univ. Press.

Irwin, Alan, and Brian Wynne. 1996. *Misunderstanding science? The public reconstruction of science and technology*. Cambridge: Cambridge Univ. Press.

Jasanoff, Sheila. 2012. *Science and public reason*. London: Routledge.

Joerges, Bernward, and Terry Shinn, eds. 2001. *Instrumentation between science, state and industry*. Dordrecht: Kluwer.

Madsen, Claus. 2010. *Scientific Europe: Policies and politics of the European Research Area*. London: Multi-Science Pub.

Nagel, Thomas. 1989. *The view from nowhere*, Oxford: Oxford Univ. Press.

Omodeo, Pietro, D. 2016. After Nikolai Bukharin: History of science and cultural hegemony at the threshold of the Cold War era. *History of the Human Sciences*, 29 (4-5): 13–34.

O'Sullivan, John. 2016. *In the footsteps of Columbus: European missions to the International Space Station*. New York: Springer.

Pollock, Ethan. 2006. *Stalin and the Soviet science wars*. Princeton: Princeton Univ. Press.

Poulantzas, Nicos. 1972. *Pouvoir politique et classes sociales*, II, Paris: Maspero.

Poulantzas, Nicos. (1978) 2000. *State, power, socialism*. London: Verso.

Salomon, Jean-Jacques. 1973. *Science and politics*. Cambridge, MA: MIT Press.

Shapin, Steven, and Simon Schaffer. 1985. *Leviathan and the air-pump. Hobbes, Boyle, and the experimental life*. Princeton: Princeton Univ. Press.

Slijper, Franck. 2005. The emerging EU military-industrial complex. Arms industry lobbying in Brussels. Transnational Institute Briefing Series 1.

Starr, Paul. 1982. *The social transformation of American medicine*. New York: Basic Books.
United Nations. 1979. *Agreement governing the activities of states on the Moon and other celestial bodies*. New York.
United Nations. 2002. *Treaties and principles on outer space*. New York: United Nations Press.
Wilson, James, Q. 1989. *Bureaucracy: What government agencies do and why they do it*. New York: Basic Books.

Acronyms

ATV	*Automated Transfer Vehicle*, ESA cargo supplier.
CADMOS	*Centre d'Aide au Développement des activités en Micropesanteur et des Opérations Spatiales. [Support to the Development of Activities under Microgravity and Space Operation Centre.]*
CNES	*Centre National d'Etudes Spatiales*, French space agency created in 1961.
ESA	*European Space Agency*, created in 1975.
ESF	*European Science Foundation*, created in 1974.
EVA	*Extravehicular activity.*
JAXA	*Japan Aerospace Exploration Agency,* created in 2003 from the fusion of ISAS (Institute of Space and Astronautical Science), NASDA (National Space Development Agency) and NAL (National Aerospace Laboratory).
MARES	*Muscle Atrophy Research and Exercise System.*
NASA	*National Aeronautics and Space Administration*, first space agency ever created (1958).
USOC	*User Support and Operation Centres*, distributed among member-states contributing to the ESA.
Roscosmos	*Russian federal space agency,* created in 1992 in the aftermath of the dismantling of the Soviet Union.

European Research Programmes
Spaces of Knowledge or Economic Tool?

5

Patricia Bauer

Abstract

The contribution discusses the development of post-Second World War (WWII) European integration and its development of an extended research policy on supranational level. In the context of research on knowledge communities it follows the question of how a specific logic of knowledge community creation has unfolded in the context of the European integration process after WWII. By applying tools from multi-level governance analysis the contribution analyses the evolvement of European research policy from EURATOM to the recent development of a "European Research Area" and an "Innovation Union". It traces the origin, direction and intention of attemps of community building in the context of European integration in order to critically discuss the hypothesis that (1) research is understood in the context of European integration as support tool for economic development, (2) that this understanding of research amplifies a specific way and quality of community building in focussing on market-relevant and exploitable research branches and (3) that the procedural guidelines for research communities delimit content, constellations and application of research outcomes. The result of the investigation supports the conjecture of a massive bias of European research policy towards economically exploitable resesarch, mainly from engeneering and natural sciences. It also proofs that community building in European research policy after WWII is mainly a top-down proactive process staged by the European Commission.

5.1 Introduction

From its beginnings, the European integration process taking place in the framework of the European Communities and the European Union has been based on the idea that communities sharing some common interest voice for an intensification of the integration, and that this interest-based deepening of integration leads to spill-overs that trigger even more integration. The neo-functionalist idea works along the line that integration processes in a delimited issue area generate payoffs for broader parts of the community and thus generate incentives for further processes of integration in other issue areas. This spill-over logic is accompanied by the creation of regimes and institutions having the capacity to oversee and regulate the complex processes created by integration. These supranational institutions become more independent and powerful as integration proceeds and are able to develop their own ideas on further steps of integration. A central assumption of neo-functionalism is the existence of elite complementarity as a precondition for spill-over and economic and political amalgamation (Haas 1958; Haas 1961; Haas 1964; Haas and Schmitter 1964).

Regarding the question whether science has established specific spaces of knowledge in Europe, the idea of elite complementary could provide an analytic framework to investigate the origin and development of European elite communities in science and academia. If we find similar patterns as Haas and colleagues identify in the economic and political integration, this would enrich our knowledge of how research communities are founded and how they are sustained. But even if the logic behind the nascency of knowledge communities is different, the rejection of neo-functionalist assumptions can help to identify the patterns and logics of community building by specifying the actual quality of knowledge and elite communities. While the case of knowledge communities in general is a wide and varied area (Kuhn 1962; Herman 1986; Kusch 2002; Lubenow 2006; see also: Gläser 2012), the subject of this contribution is the question of how a specific logic of knowledge community creation has unfolded in the context of the European integration process after the Second World War (WWII).

Since the 1980s, the enhancement of European knowledge networks and communities has become an important part of the policy-making of the European Community and, later, the European Union. The legal, financial and content-related measures contained of removing obstacles for mobility in the research sector, more and more extensive funding and enhancing European expert communities for the present leading to a future "European Research Area". The European Union's website for research and innovation confirms that research and innovation have high priority in the broad context of economic integration: "Innovation has been placed at the heart of the EU's strategy to create growth and jobs" (European Union 2016). The website further details the initiatives taken, the funding provided, and the relevant institutions and bodies documenting the extensive efforts of the European Union in this policy field. A first inspection of the policy documents reveals a certain bias in the European research policy and instructs the following multipart hypothesis: *First*, the neo-functional and market-oriented ideas and conceptions of European integration amongst policy-making elites of the European Union inspire a specific idea of research supposed to benefit the integration process as a whole. *Second*, this specific idea of research in turn supports a specific orientation of research and community building which leads to a highly selective form of research promotion. *Third*, the research promotion sets out well-defined criteria for research and knowledge communities that promote specific constellations, content and application of research. Altogether, the European Union stages research communities which serve highly specific and top-down defined objectives in accordance to the general ideas of integration and with a bias on market-oriented applicability.

This contribution first looks into the ideas and analytic concepts on the European integration process to set out the framework for the analysis of the European research policy and the discussion of the hypothesis. It then proceeds to the historical development and analyses how objectives, structures, actors, legal frameworks and funding conditions have evolved since the inception of the European integration process after WWII. It finally sketches the contours and features of the present European research policy and discusses its achievements, biases and shortcomings on the background of a simple multi-level analysis model.

5.2   Analysing the Formation of European Research Policy

For the analysis as well as the policy-making of the European Communities and the European Union, Neo-Functionalism is one of the most influential theories. The idea that integration processes in a delimited issue area generate pay-offs and thus spill-over effects on further parts of the community is supported by the practice of the early days of integrationin the first decade after the Rome Treaties of 1957. But, it is also empirically supported by a number of policy studies on the long term development of the European integration (Stone Sweet and Sandholtz 1997; Stone Sweet and Sandholtz 1998). Neo-functionalism argues that the creation of institutions is essential for processes of spill-over since the creation of new bureaucracies creates new elites identifying with their task and interpreting their role in the integration process extensively. Indeed, the policy-making in several areas of

European integration indicates that the idea of enhanced supranational authority is a highly general feature in European bodies.[1] These, again, lead to further policy initiatives in previously non-integrated areas and thus create top-down spill-overs for other issue areas leading to an intensification of integration and contributing to solidify the political process of an ever closer amalgamation of politics in Europe. Although since the 1990s intergovernmentalist (Moravcsik 1993; Puetter 2012) and constructivist (Wendt 1992; Schimmelfennig 2001) theories have challenged the validity of Neo-Functionalism for explaining European integration, bureaucratic elites continue to work on a spill-over hypothesis. The European research policy is a good example exhibiting neo-functionalist thinking in its policy documents (see below). Whereas the theoretical debates on explaining European integration have enriched the analytical tools to approach the European integration, with multi-level Governance (MLG) (Marks 1993; Hooghe and Marks 2001; Benz 2004) a more policy-oriented approach to explain the functioning, veto positions and layer-structure of European policy-making was established in the 1990s. The latter analyses European integration as multi-layered and multi-faceted processes of community-building in society and politics. Instead of the rather predictive and normatively biased neo-functionalist approach multi-level Governance offers an analytical framework for the reconstruction and explanation of how politics come about. It focusses on actors and processes of agenda setting and logics of implementation, identifying patterns of policy network formation between societal groups and policy-makers.

For identifying the patterns of European research policy-making, multi-level governance is an important tool for identifying the origin, content and function of research policy in the European integration process (see hypothesis part one). As analytical approach, multi-level Governance can help to understand which agenda-setting and implementation pathways set out by European decision makers lead to a highly selective form of research promotion and funding. This applies to the content as well as the institutional architecture of research and its funding (see hypothesis part two). Multi-level governance thus allows to identify the origin of ideas in the policy-process, the role of policy entrepreneurs in agenda-setting and

---

1   The structural similarities of policies become obvious when comparing disjunct policy areas. The regional and structural policy as one of the most powerful instruments of distribution often has served as blueprint for more recently developed policies. We can find similar logics of distribution in education policy (Bauer 1999) and the conditionality and co-financing in the Eastern enlargement and the Mediterranean policy (Tocci 2005; Zank 2005; Keck and Krause 2006; Magen 2006; Praussello 2006; Stephanou 2006; Tulmets 2006; Tulmets 2007; Gänzle 2009; European Commission and High Representative 2011; Korosteleva et al. 2014).

the creation of the research policy outfit (see hypothesis part three). Consequently, it is a promising tool to analyse the question at hand and to support our hypothetical statements on the bias of research policy community building within the European integration process.

Hence, for the question if science has established specific spaces of knowledge in the framework of the European integration process, it is important to have a closer look into the formation of European research policy. On the one hand, the framework of neo-functionalism offers an explanatory framework to analyse the objectives and aims of research policy in the general context of integration and its idea of elite complementary provides an analytic tool to explain the formation of an ever closer integration. The tools of multi-level Governance analysis allow to specify the patterns of interaction and help to identify negative (blocked decisions, incoherent agendas, inappropriate approaches, control deficits) and positive (policy learning and its conditions, policy transfer, best practices) interaction effects. Grande and Peschke (1999, 44) have argued that European research policy is confronted with a specific "linkage problem" to establish "hybrid communities" providing channels of communication, institutionalised cooperation between science, business and politics, but also linking different national research and policy-making systems. The European multi-level system thus provides an additional dimension of complexity and heterogeneity for policy-making. Taking these restrictions into account, community building in research on the European level is confronted with a rather high amount of obstacles. In particular, the independence of research from politics and business and its own logics of work processes, self organisation and interest groups together with specific national forms or organisation provide a difficult environment for an integrated European research policy.

In order to identify the origin of predominant ideas behind European community building efforts and to explain why research promotion favours specific forms, criteria, content and research constellations, it is helpful to apply the distinction of origin and rationale of interest formation processes in the European research policy arena. Following Grande and Peschke (1999, 48 ff.) we will distinguish between *bottom-up* and *top-down* processes of interest formation in order to identify the policy entrepreneurs amongst the various actors in the policy field. Bottom-up processes are forms of self-organisation of research and science sectors in Europe in order to feed European politics with expertise and to lobby for research projects. Top-down processes are characterised by the initiative of European bodies, namely the European Commission, to build a reservoir for expert advise on new programmes and to support the European bodies to persuade member states to transfer more competencies in research policy from the national to the European level. In order to clarify the rationale behind a specific initiative, programme or institution, we distinguish between *pro-active* and *reactive* processes. Whereas reactive processes of interest formation are mainly reactions of the research system on initiatives from the European political bodies that adjust to a change in circumstances for research, pro-active processes can be described as lobbying initiatives from the research system to get access to the policy formulation process on the European level.

Table 1 shows the different patterns and forms of interest formation resulting from these distinctions. It will help to identify the leading actors and the degree of autonomy behind specific policy initiatives and programmes. If most of the processes we will describe were top-down and reactive, we could conclude that the European bodies—and specifically the European Commission—were the leading policy entrepreneurs in the field. Vice versa, if we could identify pro-active bottom-up processes of policy-making, we could conclude that the European research policy originated in the autonomous interest formation of the research sectors in the European Union's / European Communities' member states.

Tab. 1   Patterns of interest formation in European research policy

	Bottom-up	Top-down
Pro-active	Transnational autonomous community building along research projects or clusters to feed the European bodies	EU bodies create expert groups for consultation on future programmes and as allies against member states' resistance
Reactive	Offices, organisations and networks to lobby for research interests in a policy initiative initiated by European bodies	EU bodies create expert groups to advise on programmes in place and for expert knowledge in selection for funding

In this scheme reactive bottom-up processes and pro-active top-down processes might be identified best with the spill-over logic of neo-functionalist thinking. In both cases, the European integration process is causative for the activation, if not creation, of interest groups and networks in the research sector. The central role of European bodies, and mainly the European Commission as a lead driver for intensified integration, would indicate how strongly content and implementation measures of European research policy are connected to other sectors and rationales of the European integration. For the field of research policy, it has already been evidenced for the case of security research (Edler and James 2015), that the Commission plays a crucial role in the agenda setting process of research policy. If we will find a strong role of European bodies, and a reactive pattern of interest formation, we best would describe the community building as a "staged" process, presumably representing the ideas and policy concepts of supranational bodies on research as a condition for economic integration, employment, growth and international competitiveness. But, if bottom-up pro-active interest formation were predominant, we could conclude on autonomous, less embedded and more heterogeneous community building.

5.3 The Beginnings of European Research Policy after WWII

European research policy after WWII is strongly connected with the reconstruction of postwar Europe and the spirit of collaboration. Its roots lie in the various attempts to establish a collaborative and competitive European industry and science sector. Thus, first attempts on a European research policy can be found in the founding statement of the Council of Europe (cit) in 1949 and the more industry- and efficiency-driven research in the framework of the European Coal and Steel Community (ECSC) founded in 1952. A major breakthrough was the establishment of CERN in 1953 to merge high energy physics research in a joint international project of twelve European countries (Papon 2004). The focus on nuclear energy in the post-world war decade followed an epistemic as well as a political motivation. In times of nuclear competition between the superpowers, it was important for the European nuclear research to keep par with the highly defense-logic driven research esp. in the US and being able to deliver world class research results. Politically, nuclear research was the key for a credible defense in the 1950s and their tense competition between the blocs in the cold war. Although the European Defense Community as a direct answer to the military threat from the communist bloc failed, the founding of the European Atomic Energy Community (EURATOM) in 1957 promised to provide a key building bloc for a future common defense. This objective, again, was followed by an intensive science and technology component of EURATOM (Guzzetti 1995; Hallonsten 2012, 301). Thus, research policy only had a limited space in the European integration project and did not spill-over from EURATOM to the areas of the economic communities, on the contrary, the crisis or EURATOM in the 1960s led to a decrease of research in the framework of the European integration process. Most relevant research at that time took place outside the framework of the Economic Communities, ECSC and EURATOM (Grande and Peschke 1999, 45). Thus, the post-world war period in Europe saw top-down initiatives, like the Council of Europe, ECSC and EURATOM to promote research as well as the very successful bottom-up research initiative of

CERN and the Conference of the European Rectors and Vice Chancellors (CRE) as interest group of European universities. The character of all these initiatives is merely pro-active because they act in order to develop the European position, be it in research, be it in the arms race.

Although EURATOM was a unique endeavour to merge nuclear research in Europe, it did not unfold impact on the broader research sector of the European Communities. On the contrary, the organisation of research in two branches, the supranationally organised Joint Research Centre and external contracts that liaised national projects with EURATOM together with a trend of re-nationalisation of politics (esp. by de Gaulle in the 1960s) led to a predominance in research and funding acquisition of the bigger countries and a debate about distribution of funding from the side of smaller countries. The 1960s thus were a decade lost for further scientific development since the nationalist bargaining structure in EURATOM prohibited a further intensification of cooperation (Guzzetti 1995). By the end of the 1960s, the debate about the "technological gap" between Europe and the US inspired new attempts in the organisation of joint European projects. By 1967 the Merger Treaty had incorporated EURATOM and ECSC with the Economic Community by integrating the steering bodies which also brought EURATOM research in closer connection with the Common Market. Together with the debate on the "technological gap" and the vivid activities to organise research outside the European integration process (mainly ESO and EMBO (Papon 2004, 64)) this created a push for a new attempt to organise European research in the framework of the EC.

The 1970s saw a major breakthrough in the development of a European research policy within the framework of the European Community. Since the legal framework for supranational competencies in research policy was limited to EURATOM, the Paris summit in 1972 decided to apply Art. 235 ECT in science and technology (Grande and Peschke 1999, 45) which widened the legal framework for EC actions. In 1974, the first *action programme* was adopted by the Council of Ministers. This consisted *sectorial programmes* on increasing competitiveness, especially aiming at IT promotion (Hallonsten 2012, 302). Whereas the European community in the early 1970s conquered research as a field relevant for the development of the common market and the competitiveness of the EC with other economic blocs, a number of bottom-up initiatives established for an intensification of pooling of funds and scale of projects. Also in 1974, national research organisations created the European Science Foundation (ESF) to broaden their collaboration. This initiative can be seen as a pro-active interest organisation of European academia in an unfolding EC policy area. The foundation of the European Space Agency (ESA) in 1975, on the other hand, is a bottom-up initiative to cope with limited national funding for space programmes and to keep par with NASA (Papon 2004, 65 f.).

Although ESA was planned to become an organisational part of the European Union, until today, it is an independent intergovernmental organisation outside the European Community/European Union.

The first half of the 1970s was formative for the foundation of European research policy because it established the way how and if the European Community/European Union is involved (top-down) and if the European bodies themselves are involved into research. The patterns established in EURATOM, the Joint Research Centre as a community-owned research institution, and indirect actions were added by concerted action:

> The Community's research activities during this period assumed three forms, which have remained practically unchanged today: direct action, indirect action, and concerted action. The first two derived from those used during the EURATOM research; the third represents a new departure. Research by direct action was carried out by the Joint Research Centre and was totally financed from the general budget of the Communities. Indirect action referred to research activities contracted out to public research centres or private laboratories in Member States; for these the Community generally paid about 50 % of the cost. In concerted actions, the Communities guaranteed and financed only the co-ordination of the research (reimbursing travel expenses, meetings etc) and the circulation of the results of the research. This last type of financing also provided an opportunity to evaluate the usefulness of individual projects which might subsequently be the object of indirect action, where this seemed to be in the Community's interests (Guzzetti 1995, 60).

The 1970 period is thus formative for the design and outfit of the European research policy of the following decades and set out what actor constellations the EC/EU would work with, what forms of coordination for research would be established and how much the EC/EU would contribute to the financing.

Tab. 2 Features of EC/EU research

	Research Actors	Mode of Coordination	Budget
Direct Action	Joint Research Centre	Advise from EC'sGD	100 % communities
Indirect Action	national public and private research institutions	Out-contracting by the EC	50 % communities / 50 % national research institutions
Concerted Action	national public and private research institutions	Network-like structure, EC as supportive actor	Only coordination and dissemination covered by the EU

5.4   The Establishment of a European Research Policy

With the project of the full establishment of the Single Market, research policy gained more importance at the beginning of the 1980s. The idea of linking technological and research progress with the competitiveness of the European Community as a whole found its expression in the sectors covered by the *First Framework Programme* (FP1) of the European Community that was adopted in 1983 and covered the period between 1984 and 1987. Industrial innovation and information technology were the key sectors the FP1 was focussing on (Guzzetti 1995, 97 ff.; Grande and Peschke 1999, 45) with energy, raw materials, environment, agriculture and industrial research as priority areas (Tamtik 2016). The new commission under Jaques Delors played a major role—amongst other great achievements in fostering the Single Market—to establish research and innovation as a community policy. With the Single European Act (SEA) a major reform of the three community treaties took place influencing content and structure of European policy-making. SEA established a community competency for science and technology and thus extended the powers given by EURATOM considerably. In perspective of integration dynamics, it changed the predominant mode of decision-making in the Council of Ministers from unanimity to qualified majority for three quarters of the decisions. With this re-arrangement, veto-players were marginalised and nationally-inspired hesitations hampering the integration process were minimised.

For the development of research policy on the European level, these steps also establish an administrative procedure to organise the policy. Art. 130i stets out that the Community's research policy will be organised in multi-annual framework programmes outlining the objectives and priorities, the financial framework and share of the Community for specific actions. Although decisions in the field of science and technology are taken unanimously by the Council, the inception of a new policy can be seen as major integration step. Unanimity still guaranteed the member states a massive influence on research policy. But with the establishment of an administrative procedure on European level, the European Commission was able to gain massive influence on the formulation of objectives and procedures of implementation. This step has been described as a shift from the "isolated" approach of EURATOM and the 1970s to a "centralised" approach that was reinforced by the Maastricht Treaty on European Union (TEU) of 1992 (Tamtik 2016). TEU stiffened the policy with rules and procedures (Banchoff 2002, 9f). Thus, we can conclude on the first generation of Framework programmes (FP1 to FP5) that the top-down approach to establish a research policy legally and administratively ended up in a quite centralised system of research funding with a central role of the European Commission. It generated a network of clientele consisting of those national research projects and institutions able to meet the Commissions criteria in the application process. Attempts to reform the system that did not connect different member states' projects failed during the 1990s due to national resistance and the resistance of the successful FP researchers (Banchoff 2002, 10ff).

It is worth to note that the thematic framework of research policy the first Framework programmes focussed only on the natural sciences. Due to its origin in EURATOM, energy research always had a prominent position, added by biotechnological and life sciences, both closely related to the supranational agricultural policy. Informations technologies and industry-related sectors added the main sectors up. Until FP4 (1994 to 199), socioeconomic research was not on the European agenda (see the excellent synopsis of Tamtik 2016, 10). Although socioeconomic research only made 1 % of all research funding, it is worth to note that its inclusion was the result of the EU's consultation with the European Science Foundation (ESF). Its focus, again, was subordinated to the aims and objectives of natural sciences research and impacts of the market integration as socioeconomic research consisted of the evaluation of science and technology policy, research on education and training and social integration and exclusion (Guzzetti 1995, 161). Attempts to reorganise the structure and reach a more comprehensive outfit of the European research policy were not successful until the end of the century, as Tamtik (2016, 12) notes:

> In 2000, Philippe Busquin launched the most ambitious effort to coordinate research policy in Europe—the European Research Area. To address the fragmentation, isolation, and compartmentalization of national research efforts and break away from the centralized approach of the Framework Programs, a change that did not involve any legal or regulatory means was needed.

5.5 After 2000: Research and Innovation as Integral Part of the EU Policy-Making

In January 2000, Commissioner Philippe Busquin launched a communication 'Towards a European Research Area' (ERA) stressing the lack of effort of the EU in research compared to the world market competitors, USA and Japan. This initiative marks the transition from a centralised to a "network-based" (Tamtik 2016) approach in European research policy which merges top-down and bottom-up modes of research organisation. This reform push for research policy, again, was rooted in economic deliberations and is embedded in the *Lisbon Strategy*. "The aim of the Lisbon Strategy, launched in March 2000 by the EU heads of state and government, was to make Europe 'the most competitive and dynamic knowledge-based economy in the world, capable of sustainable economic growth with more and better jobs and greater social cohesion'" (Committee of the Regions (CoR) 2017). In terms of policy coordination, the Lisbon Strategy introduced the Open Method of Coordination (OMC) to establish a network-based process with bottom-up and top-down processes between member states, European level and research institutions. Since the Lisbon Strategy encompassed broad areas of national authority, like employment and employability, OMC was the method to establish a durable influence structure towards the national policies.

The concept of economic growth of the Lisbon strategy consists of three pillars: innovation, learning economy and social and environmental renewal (European Council of Heads of States and Government 2000). The *European Research Area (ERA)* (for the following, see European Commission 2000), a new framing of the supranational level for improving pooling of resources and enhancement of the competitive position of the EU (Tamtik 2016), was created as integral part of the Lisbon Strategy. The ERA was defined by ten aspects (European Commission 2000, 8):

- Networking of existing centres of excellence in Europe and the creation of virtual centres through the use of new interactive communication tools.
- A common approach to the needs and means of financing large research facilities in Europe.
- More coherent implementation of national and European research activities and closer relations between the various organisations of scientific and technological cooperation in Europe.
- Better use of instruments and resources to encourage investment in research and innovation: systems of indirect aid (within the Community rules on State aid), patents, risk capital.
- Establishment of a common system of scientific and technical reference for the implementation of policies.
- More abundant and more mobile human resources:
  - Greater mobility of researchers and introduction of a European dimension to scientific careers.
  - More prominence to the place and role of women in research.
  - Stimulating young people's taste for research and careers in science.
- Greater European cohesion in research based on the best experiences of knowledge transfer at regional and local levels and on the role of the regions in the European research efforts.
- Bringing together the scientific communities, companies and researchers of Western and Eastern Europe.
- Improving the attraction of Europe for researchers from the rest of the world.

- Promotion of common social and ethical values in scientific and technological matters.

The programme of ERA in 2000 can be described as to empower the competitiveness of European research institutions by opening up new pathways of networking, collaboration, pooling of financial means and industrial investment into research on the one hand and to increase the mobility of researchers inside Europe as well as in respect of attracting researchers to Europe. The ERA represents a new quality of European research policy since its intention is the moderation and coordination of national research policies by various forms of steering. Whereas networking of national research institutions shows that the Commissions has turned its back to centralised approaches, the favouring of common approaches in financing large facilities and the idea of coherence between national and European implementation opens up new influence channels for Community ideas. The idea of knowledge transfer and the inclusion not only of research institutions but also of the regions widens the landscape of potential actors and influence points for European research policy. Since the ERA was further on staffed with the funding instruments of the Framework programmes, the financial incentives of the FPs were targeting not only research institutions in the member states but now embraced also the regions as actors. This move can be explained by the positive experience of decades of regional and cohesion policy that established a triangle structure between European, national and regional level that often was able to sandwich national hesitations or opposition by a strong alliance between the European Commission and the European regions, not least established by strong funding incentives from the European side (for this argument, see Lang and Heinelt 2011).

Together with soft coordination mechanisms like benchmarking and best practice policy-learning within OMC and the less aggressive economic integration framing, the widening of the actor structure gave the Commission a decisive role in collecting and filtering information. Additionally, the Framework Programmes in the hands of the Commission could set strong incentives in content as well as structures of the successful projects. ERA thus developed from a vague idea of a common market for research into a multi-layered system that was able to harmonise amongst different national research systems and to assure high compliance of national research actors with ERA (Lepori et al. 2014). ERA thus represents a top-down policy approach to open-up new bottom-up and network channels of policy learning for a proactive policy of economic growth by the European level. This soft steering approach nevertheless put the Commission into the centre for control due to the strong funding incentives of the FPs.

Although with these arrangements the scene was set to unfold a common trans-national research policy in the EU, the Lisbon strategy and ERA witnessed a major setback due to the financial crisis. Taking into consideration that re-arrangements of policy structures take considerable periods of transition, in particular if larger parts of national policies are affected, the High Level Group chaired by Wim Kok presented a very disenchanting judgement of the progress of Lisbon at the end of 2004:

> External events since 2000 have not helped achieving the objectives but the European Union and its Members States have clearly themselves contributed to slow progress by failing to act on much of the Lisbon strategy with sufficient urgency. This disappointing delivery is due to an overloaded agenda, poor coordination and conflicting priorities. Still, a key issue has been the lack of determined political action. The Lisbon strategy is even more urgent today as the growth gap with North America and Asia has widened, while Europe must meet the combined challenges of low population growth and ageing. Time is running out and there can be no room for complacency. Better implementation is needed now to make up for lost time (Kok 2004, 6).

As a consequence, in early 2005 the Commission suggested to the Spring European Council to relaunch the Lisbon Strategy by improving implementation and regulatory frameworks. Additionally, the development of a knowledge society, the crucial role of research for the European Union's position on world markets and efforts towards innovation are the thematic issues focussed on. It is up to the discussion if these attempts to re-arrange the Lisbon Strategy were not efficient or if the World Economic Crisis from 2007 on was the main factor of failure (see Lucian 2015). However, by 2010 the European Union created a new strategy, *Europe 2020*, for the period from 2010 to 2020 to initiate "smart, sustainable, inclusive growth" (European Commission 2010a, 10; Wandel 2016) Improvement of economic performance is put in concrete terms in the strategy document. The Commission proposes headline targets focussing on the economic core of the European integration:

- 75 % of the population aged 20-64 should be employed
- 3 % of the EU's GDP should be invested in R&D.
- The "20/20/20" climate/energy targets should be met (including an increase to 30 % of emissions reduction if the conditions are right).
- The share of early school leavers should be under 10 % and at least 40 % of the younger generation should have a tertiary degree.
- 20 million less people should be at risk of poverty (European Commission 2010a, 5)

These objectives build the framework of the activities carried out by the EU but also demanded from the member states. In this context, the Commission proposed seven flagship initiatives "to catalyse progress under each priority theme:

- "Innovation Union" to improve framework conditions and access to finance for research and innovation so as to ensure that innovative ideas can be turned into products and services that create growth and jobs.
- "Youth on the move" to enhance the performance of education systems and to facilitate the entry of young people to the labour market.
- "A digital agenda for Europe" to speed up the roll-out of high-speed internet and reap the benefits of a digital single market for households and firms.
- "Resource efficient Europe" to help decouple economic growth from the use of resources, support the shift towards a low carbon economy, increase the use of renewable energy sources, modernise our transport sector and promote energy efficiency.
- "An industrial policy for the globalisation era" to improve the business environment, notably for SMEs, and to support the development of a strong and sustainable industrial base able to compete globally.
- "An agenda for new skills and jobs" to modernise labour markets and empower people by developing their of skills throughout the lifecycle with a view to increase labour participation and better match labour supply and demand, including through labour mobility.
- "European platform against poverty" to ensure social and territorial cohesion such that the benefits of growth and jobs are widely shared and people experiencing poverty and social exclusion are enabled to live in dignity and take an active part in society." (European Commission 2010a, 5–6)

Europe 2020 models economic growth of the European Union along the the ideas of employment, energy saving, ecology and anti-poverty and educational strategies. It merges more traditional ideas of industrial policies supporting economic growth with progress oriented topics like energy and educational training. It is remarkable that the "Innovation Union" initiative now openly states that research and innovation should be turned into products that create jobs. The sub-ordination of research and innovation under economic growth is further spelled out in the concept of the "Innovation Union". This consists of mainly practically-oriented thematic research fields that should be connected to the problems of ageing society and ecological sustainability the EU is facing. Whereas on European level the European Union requests to complete the ERA and to mobilise funding also from the regional and cohesion policy, it attempts also to mainstream national research policies.

The latter consists of a number of measures to further transnational research but also to utilise the national education and training systems to "ensure a sufficient supply of science, maths and engineering graduates and to focus school curricula on creativity, innovation, and entrepreneurship" (European Commission 2010a, 13). This deep intervention into national systems is amplified by the "Youth on the Move" initiative that requests a modernisation of higher education as well as a more labour-market oriented education policy that improved educational outcomes on all stages—including pre-school and primary sectors.

The new subjection of education and science under economic duties and constraints is accompanied by a regional policy re-orientation towards "smart growth in Europe" in October 2010 (European Commission 2010b). Although regional policy was already included under ERA into the framework of research policy, this step consists a further focussing of regions as partners for the Innovation Union[2] by enforcing ERDF funding for education, research and innovation. In particular, interregional cooperation, co-financing of FP7 projects and projects of innovative character are highlighted to be prioritised and funded by ERDF. Since the regional policy is the biggest supranational means in economic policy, this move implies a major shift in the alignment of European policies. A strict delimitation of regional, educational and research policy seems to be no longer intended which is also represented by the inclusion of regional policy on the webpage of GD Research and innovation, titled: "synergies with regional funds" (European Commission GD Research and Innovation 2017). Whereas the inclusion of regional policy into the innovation objectives of European economic policy has fundamentally blurred the edges of research, the complex network structure of today's European research policy is complemented by a new Framework Programme covering the period between 2014 and 2020; this 8th FP is *Horizon 2020*.

The programme establishing document concedes that "Horizon 2020 will be a key tool in implementing the Innovation Union flagship initiative, in delivering on the commitments made therein ..." (European Commission 2011, 2) Horizon 2020 is pooling the funding for research, innovation related fields of the Competitiveness and Innovation Framework Programme and the European Institute of Innovation and Technology (EIT). The latter was established in Budapest in 2008 and was intended to contribute to the establishment of "knowledge and innovation communities". It works as a network organiser for currently six knowledge and innovation communities consisting of private companies, public institutions and

---

2   On the conceptual background of the relevance of regional policy for innovation see: Camagni and Capello (2013).

research institutions (for further details see: European Institute of Innovation and Technology 2017). The thematic focus of Horizon 2020 is on three pillars:

- "Excellent Science" consisting of European Research Council and Marie Sklo-dowska-Curie Action
- "Industrial Leadership" consisting of research fora on information and communication technologies, nanotechnologies, advanced materials, advanced manufacturing and processing, biotechnology, space
- "Societal Solutions" consisting of applied solutions for problems as health, energy, security, transport, climate, European society (European Commission 2011, 4ff.)

With exception of the "excellent science" pillar, all activities within Horizon 2020 focus on applied sciences convenient for the overarching objective of an innovative market economy in Europe that is able to challenge Japan and the US on the world market. It is worth to mention that basic research is marginalised in the "excellence" pillar. Research on, lets say, fundamental philosophical questions, sociological theorising, critical security studies in political science or linguistics have rather limited access to European research funding and play a marginal role in the ideas about a European Research Area. Horizon 2020 together with the project of the European Research Area and the Innovation Union manifest the concept of research as a precondition for economic competitiveness. Moreover, the economic fixation in European research policy neglects the role of basic research for knowledge production and as fertile ground for technological innovation at all. This is well illustrated by an image of a "knowledge triangle" as promoted by the EIT, where research and technology form one vertex of a triangle along with higher education and business, and where all vertices are seen to contribute equally to the evolvement of entrepreneurship (see European Institute of Innovation and Technology 2017).

The European Union's approach to research as exploitable findings for innovation, economic growth and competitiveness has been moderated recently by a growing weight of "responsible research and innovation" in order to engage the public sector in responding to the needs of the citizens of Europe, including innovative health and environmental research (Tamtik 2016, 16). This can be interpreted as an attempt of the Commission to create a higher legitimation for its approach by including the interests of the citizens into the market-driven economic agenda. In principle, it does not change the quality of research and innovation policy of today as an applied natural sciences funding machine.

5.6   Conclusion: The Anatomy of European Research Policy

We have analysed the evolvement of a European research policy from its beginnings in EURATOM until today's complex Innovation Union structure. Although we can observe vivid bottom-up activities of autonomous organisation of research in Europe—especially in the beginnings after WWII—the dominant pattern of integration and community-building is an intentional public sector engagement and active promotion of closer cooperation. Especially the High Authority (of EURATOM) and the European Commission play a key role in the institutional network of European research policy by setting and shaping research objectives. Whereas in the early stages of European Research policy influential interest groups were able to establish an alternative research space, pro-active bottom-up activities diminish the more active the European Commissions role develops. This is accompanied and bolstered by the unfolding of increasing funding opportunities from the European level. The funding incentives again allowed to establish specific constitutionalities for projects that required partnering, trans-nationalisation and in the years from 2000 on the inclusion of regional development and business partners. The more and more complex structure of research policy since 2000 and the consciously blurring of economic competitiveness, education, innovation and research created, together with OMC steering mechanisms network-like structures in research policy. Nevertheless, the European Commission by setting the procedural framework and defining the priority areas of research was at the centre and the key actor to push research policy into the direction of applied sciences. Research policy and research communities in Europe can thus be characterised as majorly top-down staged processes.

Whereas the European Commission constructed over time a complex multi-level, multi-sector system of policy networks in research policy, it is remarkable how unchanged the functional role of research has been over time. The neo-functionalist logic is still well recognisable in the policy documents over time ascribing research a key role for the future global economic position of the EU. Form EURATOM days on, the idea of integrating research into economic strategies is part of the European research policy discourse. It is not only since the last round of reshuffling with ERA, Europe 2020 and Horizon 2020 that its role is procedurally as well as in its strategic importance for the whole of the European integration process massively upgraded.

Although research is of central importance for the European Union, bottom-up interest formation for basic research, the arts and the humanities was remarkably unsuccessful. The bias on applied sciences from its beginnings has developed in a more gentle formulated preference for business, employment and growth relevant innovation. As the role of the European research policy and the financial incentives grew, the role of the European Commission as agenda-setting body and moderator in complex research-business networks became of decisive importance. Thus, it is the ideas of the Commission of how research is related to integration that establish the framework of research activities.

We have argued that the central idea on research as tool for integration has only changed marginally over time. Nevertheless, the evolution from EURATOM to a complex innovation Union was also accompanied by changes in discursive frameworks. The permanent discursive recalibration has contributed to fudging the delimitations between sectors and policy areas (research, innovation knowledge society, mobility). This has, until today, created a non-discernibility of research from other economic processes. The discursive dominance of growth (in the context of globalisation), innovation and employability has legitimised the preferences of the European level for application oriented research. Moreover, the creation of the multifold research and knowledge communities which have benefitted from the European funding has an amplifying effect on the content and direction of European research policy. Since the Commission manly relies on experts formerly or still involved in EU supported projects, uncritical and affirmative consultancy may lead to a vicious circle. Critical reflections on the impact of the Commission's strategy as a whole are thus sidelined by the logic of research funding.

We had established a multi-part hypothesis on European research policies and can state that we first can support that neo-functional and market-oriented ideas and conceptions of European integration amongst policy-making elites of the European Union—especially the European Commission as key agenda setting body—inspire the idea that research can support the integration process as a whole. *Second*, we can affirm that the top-down mode of steering is—specifically since the 1980s—massively shaping what is understood as relevant research in the European Community / European Union. We have to concede that since the 2000s more complex network-like structures are prevalent. But, the more actors and relevant sectors involved, the more relevant become agenda-setting and moderating actors. In this role, the European Commission has even gained importance and has thus marginalised autonomous knowledge communities and alternative ideas on the role of research in Europe. Thus, basic research in the humanities and the arts only finds its place in European research policy if funded by the excellence pillar. *Third*, basic research hardly finds a place in the European Research Area due to the lack of exploitability and relevance for the economic development of the Union. We can thus support that the European Union stages research communities and networks which serve very specific and top-down defined objectives in accordance to the general neo-functionalist ideas of integration and with a bias on market-oriented applicability. The system of multi-level governance serves in this context as a method of solving complex economic growth problems and to legitimise the mainstreaming of research systems in Europe with economic necessities, consumer relevance, increase of employment and effectiveness.

References

Banchoff, Thomas. 2002. Institutions, inertia and European Union research policy. *Journal of Common Market Studies* 40: 1–21. doi:10.1111/1468-5965.00341.

Bauer, Patricia. 1999. *Europäische Integration und deutscher Föderalismus. Eine Untersuchung des europäischen Mehrebenenregierens im Bildungsbereich*. Münster: agenda.

Benz, Arthur. 2004. Multilevel Governance – Governance in Mehrebenensystemen. In *Governance – Regieren in komplexen Regelsystemen: Eine Einführung*, ed. Arthur Benz, 125–146. Wiesbaden: VS Verlag für Sozialwissenschaften.

Camagni, Roberto, and Roberta Capello. 2013. Regional innovation patterns and the EU regional policy reform: Toward smart innovation policies. *Growth and Change* 44: 355–389. doi:10.1111/grow.12012.

Committee of the Regions (CoR). 2017. CoR – The Lisbon Strategy in short. *Europe 2020 Monitoring Platform*. https://portal.cor.europa.eu/europe2020/Profiles/Pages/TheLisbon-Strategyinshort.aspx. Accessed July 18.

Edler, Jakob, and Andrew D. James. 2015. Understanding the emergence of new science and technology policies: Policy entrepreneurship, agenda setting and the development of the European Framework Programme. *Research Policy* 44: 1252–1265. doi:10.1016/j.respol.2014.12.008.

European Commission. 2000. *Towards a European Research Area*. Communication from the Commission to the European Parliament, the Council, the European Economic and Social Committee and the Committee of the Regions. COM(2000) 6. Brussels.

European Commission. 2010a. *Europe 2020: A strategy for smart, sustainable and inclusive growth*. Communication from the Commission COM(2010) 2020. Brussels: European Commission.

European Commission. 2010b. *Regional Policy Contributing to smart growth in Europe*. Communication from the Commission to the European Parliament, the Council, the Economic and Social Committee and the Committee of the Regions COM(2010) 553. Brussels: European Commission.

European Commission. 2011. *Horizon 2020: The framework programme for research and innovation*. Communication from the Commission to the European Parliament, the Council, the European Economic and Social Committee and the Committee of the Regions COM(2011) 808 final. Brussels: European Commission.

European Commission GD Research and Innovation. 2017. Synergies with Structural Funds.

European Commission, and High Representative. 2011. *Implementation of the European Neighbourhood Policy in 2010: Follow-up to the Joint communication on a partnership for democracy and shared prosperity with the southern Mediterranean*. Joint Staff Working Paper SEC(2011) 638. Brussels: European Commission.

European Council of Heads of States and Government. 2000. Presidency Conclusions of the Lisbon European Council. March 23.

European Institute of Innovation and Technology. 2017. EIT – Making innovation happen. *European Institute of Innovation & Technology (EIT)*. https://eit.europa.eu/eit-home. Accessed July 20.

European Union. 2016. European Union – Research and Innovation. Text. *European Union – European Commission*. June 16.

Gänzle, Stefan. 2009. EU governance and the European Neighbourhood Policy: A framework for analysis. *Europe-Asia Studies* 61: 1715–1734. doi:10.1080/09668130903278926.

Gläser, Jochen. 2012. Scientific communities. In *Handbuch Wissenschaftssoziologie*, 151–162. Wiesbaden: Springer VS.

Grande, Edgar, and Anke Peschke. 1999. Transnational cooperation and policy networks in European science policy-making. *Research Policy* 28: 43–61. doi:10.1016/S0048-7333(98)00099-7.

Guzzetti, Luca. 1995. *A brief history of European Union research policy*. European Commission Science Research and Development Studies 5. Luxembourg: Office for Official Publications of the European Communities.

Haas, Ernst B. 1958. *The uniting of Europe: Political, social and economic forces 1950–1957*. London: Stevens.

Haas, Ernst B. 1961. International integration: The European and the universal process. *International Organization* 15: 366–392. doi:10.1017/S0020818300002198.

Haas, Ernst B. 1964. *Beyond the nation-state: Functionalism and international organization*. Stanford, CA: Stanford Univ. Press.

Haas, Ernst B., and Philippe C. Schmitter. 1964. Economics and differential patterns of political integration: Projections about unity in Latin America. *International Organization* 18: 705–737. doi:10.1017/s0020818300025297.

Hallonsten, Olof. 2012. Continuity and change in the politics of European scientific collaboration. *Journal of Contemporary European Research* 8: 300-319.

Herman, Ros. 1986. *The European scientific community*. Harlow: Longman.

Hooghe, Liesbet, and Gary Marks. 2001. *Multi-level governance and European integration*. Lanham, MD: Rowman & Littlefield.

Keck, Wolfgang, and Peter Krause. 2006. *How does EU enlargement affect social cohesion?* Discussion papers 601. Berlin: German Institute for Economic Research, DIW.

Kok, Wim. 2004. *Facing the challenge: The Lisbon Strategy for Growth and Employment*. Report from the High Level Group. Luxembourg: Office for Official Publications of the European Communities.

Korosteleva, Elena, Michal Natorski, and Licínia Simão, ed. 2014. *EU policies in the Eastern neighbourhood: The practices perspective*. London: Routledge.

Kuhn, Thomas S. 1962. *The structure of scientific revolutions*. Chicago: Univ. of Chicago Press.

Kusch, Martin. 2002. *Knowledge by agreement: The programme of communitarian epistemology*. Oxford: Clarendon Press.

Lang, Achim, and Hubert Heinelt. 2011. Regional actor constellations in EU Cohesion Policy: Differentiation along the policy cycle. *Central European Journal of Public Policy* 5: 4–28.

Lepori, Benedetto, Emanuela Reale, and Philippe Larédo. 2014. Logics of integration and actors' strategies in European joint programs. *Research Policy* 43: 391–402. doi:10.1016/j.respol.2013.10.012.

Lubenow, W. C. 2006. Knowledge communities in Europe from the Renaissance through the Cold War. *Interdisciplinary Science Reviews* 31: 105–120. doi:10.1179/030801806X103334.

Lucian, Paul. 2015. From the Lisbon Strategy to Europe 2020. *Studies in Business and Economics* 10: 53–61. doi:10.1515/sbe-2015-0020.

Magen, Amichai. 2006. The shadow of enlargement: Can the European Union Neighbourhood Policy achieve compliance? *Columbia Journal of European Law* 12: 495–538.

Marks, Gary. 1993. Structural policy and multilevel governance in the EC. In *The Maastricht debates and beyond*, ed. Alan W. Cafruny and Glenda Rosenthal, 391–410. Boulder, CO: Lynne Rienner.

Moravcsik, Andrew. 1993. Preferences and power in the European Community: A liberal intergovernmentalist approach. *Journal of Common Market Studies* 31: 473–524. doi:10.1111/j.1468-5965.1993.tb00477.x.

Papon, Pierre. 2004. European scientific cooperation and research infrastructures: Past tendencies and future prospects. *Minerva* 42: 61–76. doi:10.1023/B:MINE.0000017700.63978.4a.

Praussello, Franco. 2006. *Sustainable development and adjustment in the Mediterranean countries following the EU enlargement*. Milano: F. Angeli.

Puetter, Uwe. 2012. Europe's deliberative intergovernmentalism: The role of the Council and European Council in EU economic governance. *Journal of European Public Policy* 19: 161–178. doi:10.1080/13501763.2011.609743.

Schimmelfennig, Frank. 2001. The community trap: Liberal norms, rhetorical action and the eastern enlargement of the European Union. *International Organization* 55: 47–80. doi:10.1162/002081801551414.

Stephanou, Constantine A., ed. 2006. *Adjusting to EU enlargement: Recurring issues in a new setting*. Cheltenham: Edward Elgar.

Stone Sweet, Alec, and Wayne Sandholtz. 1997. European integration and supranational governance. *Journal of European Public Policy* 4: 297–317. doi:10.1080/13501769780000011.

Stone Sweet, Alec, and Wayne Sandholtz. 1998. Integration, supranational governance, and the institutionalization of the European polity. In *European integration and supranational governance*, ed. Wayne Sandholtz and Alec Stone Sweet, 1–26. Oxford: Oxford Univ. Press. doi:10.1093/0198294646.003.0001.

Tamtik, Merli. 2016. Institutional change through policy learning: the case of the European Commission and research policy. *Review of Policy Research* 33: 5–21. doi:10.1111/ropr.12156.

Tocci, Nathalie. 2005. Does the ENP respond to the EU's post-enlargement challenges? *International Spectator* 40: 21–32. doi:10.1080/03932720508457108.

Tulmets, Elsa. 2006. Adapting the experience of enlargement to the neighbourhood policy: The ENP as a substitute to enlargement? In *The EU and its neighbourhood: Policies, problems, priorities*, ed. Petr Kratochvíl, 29–57. Prague: Institute of International Relations.

Tulmets, Elsa. 2007. Policy adaptation from the enlargement to the neighbourhood policy: A way to improve the EU's external capabilities. In *Sécurité extérieure de l'UE: Nouveaux territoires, nouveaux enjeux*, ed. Sandra Lavenex and Frédéric Mérand, 55–80. Paris: L'Harmattan.

Wandel, Jürgen. 2016. The role of government and markets in the strategy "Europe 2020" of the European Union: A robust political economy analysis. *International Journal of Management and Economics* 49: 7–33. doi:10.1515/ijme-2016-0002.

Wendt, Alexander. 1992. Anarchy is what states make of it: The social construction of power politics. *International Organization* 46: 391–425. doi:10.1017/s0020818300027764.

Zank, Wolfgang. 2005. *The politics of eastern enlargement: Historical reconstruction and theoretical conclusions*. Occasional Papers 38. Aalborg: European Research Unit, Aalborg University.

Expert Communities and Security

Neofunctional Integration in European Policing Practices

6

Stephen Rozée

Abstract

This chapter provides an examination of European integration in areas of policing practices from a neofunctionalist perspective. It is argued that expert communities, knowledge exchange, and increasing common practices in policing have developed in response to other areas of European integration. Activities in both the internal (within the European Union (EU)) and external (outside of the EU) dimensions of European policing are considered, with particular focus on the European Union Agency for Law Enforcement Cooperation (Europol) and the European Union Rule of Law Mission in Kosovo (EULEX). This chapter argues that, while neofunctionalism is able to provide some insights into the development of European integration in policing, differing national priorities and expert police practitioners' lack of commitment to EU policing mechanisms and activities have sometimes formed obstacles to effective cooperation.

Following the Treaty of Maastricht coming into effect in 1993, the European Union (EU) has developed a range of instruments aiming to increase police cooperation in the fight against organized crime and terrorism. Key official documents on the EU's security strategy, such as the 2004 Hague and 2010 Stockholm Programmes, emphasise the importance of the European Police Office (Europol) and EU civilian missions carried out in post-conflict countries in the EU's neighbourhood for addressing major threats to European security. Furthermore, major terrorist attacks occurring in EU member states have placed police cooperation in counter-terrorism high on the EU's security agenda.

This chapter presents an examination of European integration in key areas of policing from a neofunctionalist perspective. It is argued that expert communities, knowledge exchange, and increasing areas of common practices in policing have developed in response to other areas of European integration. Activities in both the internal dimension (within the EU's geographical area) and external dimension (outside of the EU) of European policing are considered. Particular attention is given to Europol, the EU police agency, including its information and intelligence-sharing role as well as its involvement in joint operations among EU member states' police agencies, and to the EU's Rule of Law Mission in Kosovo, which is the most extensive EU civilian mission to date. This chapter seeks to address two questions: 1) To what extent have European expert communities formed in policing? 2) How can a neofunctionalist perspective explain European integration in policing activities and the formation of expert communities in policing in the EU?

The structure of the chapter is as follows. First, an explanation will be given of the neofunctionalist view of European integration; second, European integration in policing and the formation of expert policing communities will be examined from a neofunctional perspective, focusing on the case of Europol; third, a neofunctionalist analysis of the EU Rule of Law Mission in Kosovo will be presented; finally, the conclusion section will offer overall findings regarding the strengths and limitations of neofunctionalism for explaining European integration in policing.

6.1   European Integration and Neofunctionalism

Neofunctionalism has become a well-established theory for explaining and understanding European Integration, as well as a dominant perspective in European Union studies. As Rosamond (2000, 50) states, "For many, 'integration theory' and 'neofunctionalism' are virtual synonyms". Neofunctionalism builds on the work of scholars such as David Mitrany, who had argued that the design of authoritative institutions would be determined by the "functions" (that is, fulfillment of particular needs) they were required to perform; international agencies would be more efficient at meeting certain functions than national governments, and this would lead to areas of interdependence and integration among states (Rosamond 2000, 31–38; McCormick 2014, 9). These ideas were expanded as neofunctionalism by Ernst Haas in 1958, whose book *The Uniting of Europe: Political, Social, and Economic Forces, 1950–1957* provided the first in-depth study of European integration (McCormick 2014, 8). In this work, Haas defined integration as:

The process whereby political actors in several distinct national settings are persuaded to shift their loyalties, expectations and political activities toward a new centre, whose institutions, possess or demand jurisdiction over the pre-existing national states. The result of a process of political integration is a new political community, superimposed over the pre-existing ones. (Haas 1958, 16)

Lindberg builds on Haas's neofunctionalist definition of integration as follows, emphasizing the idea that integration involves nations neither wishing nor being able to conduct policies independently of each other:

1. The process whereby nations forego the desire and ability to conduct foreign and domestic policies independently of each other, seeking instead to make joint decisions or to delegate the decision-making process to new central organs: and (2) the process whereby political actors in several distinct settings are persuaded to shift their loyalties and political activities to a new centre. (Lindberg 1963, 9)

This neofunctionalist understanding of integration proposed by Haas and Lindberg highlights the move away from nationalism to cooperation. However, certain pre-conditions are required for integration to occur: first, there must be a switch within the public away from nationalist attitudes and towards international cooperation; second, a desire among elites to promote integration is necessary—this should be for *pragmatic* reasons, rather than as a result of altruistic motivations; and third, real power should be delegated to a new supranational authority (Rosamond 2000, 31–38; McCormick 2014, 8). An expansion of integration would take place once these changes occurred, caused by one of the most important neofunctionalist concepts: the notion of "spillover" (Rosamond 2000, 58).

According to Lindberg (1963, 10; also McCormick 2014, 8), spillover is the process by which "a given action, related to a specific goal, creates a situation in which the original goal can be assured only by taking further actions, which in turn create a further condition and a need for more action". McCormick (2014, 8) gives the example of integration in agriculture, explaining that, for integration to work effectively in that area, integration in the areas of transport and agricultural support services would also be needed; in turn, the integration of an international rail system may lead to the requirement of integration of road and air travel routes. The emergence of various areas of integration seen in the European Union may be explained through the neofunctionalist perspective that Haas and Lindberg proposed, and the occurrence of integration in an increasing number of areas can understood as a result of the process of spillover. This chapter will later explore examples of this in European police activities. Spillover may be divided into three main categories:

- *Functional spillover.* This is when states' economies are interconnected to the extent that if they integrate one sector, this will lead to integration in other sectors. This process would build many links among states and would require the creation of a large number of functional bodies, leading to a decline in the power of national governmental institutions; this will drive forward economic and political integration.
- *Technical spillover.* This occurs when differences in standards among states cause regulations to rise or sink to level of the state with the highest or lowest regulations. An example of this in the EU is environmental regulation, where Bulgaria and Romania adopted higher standards of environmental controls to comply with EU laws—laws that had been driven by pressures from EU member states with high levels of environmental controls, including Germany and the Netherlands.
- *Political spillover.* This occurs when particular functional sectors have been integrated and interest groups (such as labour unions and lobbying groups) move to trying to influence supranational institutions instead of national governments. Supranational institutions would encourage this switch in order to try to acquire increased powers for themselves, and interest groups would benefit from integration and form a barrier against retreating from integration. In this way, politics would increasingly move from the national level to take place at the supranational level. (Bache et al. 2011, 10, quoted in McCormick 2014, 10)

According to McCormick (2014, 11), the EU has always demonstrated relatively high potential for integration, due to its member states being economically compatible in most respects and having broadly similar interests and goals, and because interest groups have benefitted from the formation of the EU level of decision-making. The development of the European Union has not been characterized not only by economic integration; areas of political integration have also been established through the EU's treaties. Integration in areas relating to security, particularly police and judicial cooperation, has developed rapidly since the Maastricht Treaty came into effect in 1993, and has continued to increase significantly under the 2009 Lisbon Treaty. The next section of this chapter provides a brief outline of the development of EU policies in police cooperation. Following this, EU police cooperation and the formation of expert communities in European police activities will be examined from a neofunctionalist perspective.

6.2 Expert Communities in Policing – the Development of Police and Judicial Cooperation in the European Union

Since the Maastricht Treaty came into effect in 1993, police and judicial cooperation has increased significantly within the EU's Area of Freedom, Security and Justice (AFSJ). As counter-terrorism and combating organised crime in Europe have become political priorities, so the AFSJ has become one of the most rapidly developing areas of integration in the European Union (Kaunert et al. 2014).

Concerns relating to European security have driven forward developments in the AFSJ, resulting in many new police cooperation policies and instruments being produced by the EU (Bigo et al. 2010; Rozée 2013, 2015; Kaunert et al. 2014). Examples of EU policing activities include Europol, the EU policing agency that supports the law enforcement agencies of member states; databases containing information that can assist police investigations, such as the Schengen Information System (SIS), customs cooperation centres, and EU civilian missions that are undertaken around the world to provide policing services and training for local police in post-conflict areas. While some limited intergovernmental cooperation had existed prior to 1993, such as within the Trevi group, the Maastricht Treaty significantly built on this, giving the EU its first competences in internal security and justice matters (Kaunert et al. 2014). EU justice and security cooperation increased in prominence with he 1999 Treaty of Amsterdam, which re-labelled Justice and Home Affairs (JHA) policies as the "Area of Freedom, Security and Justice". In 2009, the Lisbon Treaty abolished the EU's "three pillar" structure (the three pillars being "European Communities", "Common Foreign and Security Policy (CFSP)", and "Police and Judicial Cooperation in Criminal Matters"), increasing supranational competences in criminal justice and policing matters (Kaunert 2010).

This chapter will examine two of the most important examples of the EU's involvement with policing activities, particularly focusing on where expert communities have formed as part of those activities. These examples will be analysed from a neofunctionalist perspective, in order to determine the extent to which neofunctionalism can explain European integration in policing activities in these cases, the formation of European expert communities in policing, and to highlight some of the successes and shortcomings of EU integration in the field of policing. The examples included are: (1) Europol, the law enforcement agency of the European Union and the EU's main instrument for police cooperation; (2) The European Union Rule of Law Mission in Kosovo (EULEX), the most extensive of the EU's civilian missions in terms of the policing activities that it contains within its mandate.

6.3   The Origin and Tasks of Europol

The European Police Office (Europol) is a European Union law enforcement agency that aims to improve effectiveness and cooperation among EU Member States' competent authorities in preventing and combating terrorism, drug trafficking, and other types of organised crime, and it is the EU's main instrument for police cooperation. Europol was first established under the 1992 Maastricht Treaty. While Europol begun operations in a limited capacity in 1994 as the Europol Drugs Unit, the agency's original legal basis was formally set out in the 1995 Europol Convention. It took until July 1999 for every EU Member State to ratify the Convention, and it was at that point that Europol was able to begin its full activities (Bures 2008, 501; Kaunert, 2010, 653–4). Europol's headquarters is situated in The Hague, and as of 2016 over 1000 members of staff are based there (Europol 2016).

Europol does not operate as an executive police force, as it has no direct powers to launch its own investigations or make arrests. Instead, it offers support to Member States' law enforcement agencies by gathering, analysing and disseminating information and coordinating operations (Europol 2016). In addition, Europol assists with the investigation of criminal cases that occur in EU countries by providing experts and analysts taking part in Joint Investigation Teams (JIT). Furthermore, Europol has formed operational and strategic agreements with many third states and non-EU institutions, including Australia, Canada, Croatia, Iceland, Norway, Switzerland, and the United States of America, as well as several such as Interpol and the United Nations Office on Drugs and Crime (Kaunert 2010, 655; Europol 2016).

The formal activities of Europol were set out in Article 3 of the 1995 Europol Convention as follows: (1) to facilitate the exchange of information between the Member States; (2) to obtain, collate and analyse information and intelligence; (3) to notify the competent authorities of the Member States without delay via the national units referred to in Article 4 of information concerning them and of any connections identified between criminal offences; (4) to aid investigations in the Member States by forwarding all relevant information to the national units; (5) to maintain a computerised system of collected information containing data; (6) to participate in a support capacity in Joint Investigation teams; and (7) to ask the competent authorities of the Member States concerned to conduct or coordinate investigations in specific cases. In addition, Article 3(2) of the Europol Convention sets out further tasks, which includes the development of specialist knowledge of the investigative procedures of Member States' law enforcement authorities, producing situational reports and providing strategic intelligence—a key example of the latter is the annual EU Serious and Organised Crime Threat Assessment report (SOCTA) that Europol produces to assist EU member states' law enforcement authorities (Rozée et al. 2014). The following section of this chapter will examine how these tasks have been carried out through the formation of European expert communities in policing.

6.4   Europol and Expert Communities in Policing

There are a number of areas of Europol's activities that have led to the formation of European expert communities in policing. First, Europol's staff in the Hague, numbering over 1,000 individuals, consists of analysts, linguists, information and communications technology experts, experienced police personnel and other specialists from around Europe (Europol 2017). Europol functions by bringing these experts into a working environment together, where their expertise is coordinated to address security and crime issues as diverse as terrorism, drug trafficking, money laundering, organised fraud and people smuggling (Bures 2008; Rozée et al. 2014; Europol 2017). It can therefore been seen that Europol essentially functions as an expert policing community, drawing together experience and knowledge from around Europe.

Second, Europol operates by working closely with the law enforcement agencies of EU member states, creating an interconnected network of expert police practitioners from across the whole of the EU. This includes information and intelligence-sharing among Europol and member states' law enforcement agencies, as well as analytical and strategic support being provided by Europol. The primary way this relationship functions is through the Europol National Units (ENU) and liaison officers based in Europol's headquarters in The Hague. Each EU member state has a designated ENU acting as a liaising body between Europol and the competent authorities of that member state. At least one liaison officer to be based at the Europol headquarters is seconded to Europol by each national unit (Kaunert 2010; Rozée et al. 2014). While the liaison officers' role is to represent their national unit's interests, a collaborative network of European expert policing practitioners based in Europol's headquarters and in each EU member state has been formed (Kaunert 2010; Rozée et al. 2014).

Third, the major databases that Europol maintains as part of the Europol Computer System, which are among the most important of Europol's activities, provide a further example of European expert communities in policing. Europol's databases include the European Information System (EIS), containing data on individuals suspected of serious criminal activities within Europol's remit, and the Analysis Work Files (AWF), which provide intelligence analysis to support investigations being carried out by Member States' law enforcement agencies involving organised crime and terrorism (Europol 2006, 2–6; Rozée et al. 2014). The creation of AWFs provides an example of European expert communities being formed; these databases are created by the collaborative efforts of Europol's experts at the Hague as well as national experts who are seconded to Europol (Occhipinti 2003, 61; Kaunert 2010, 655).

Having provided the above examples of Europol being involved in the formation of European expert communities in policing, the following section of this chapter will examine the extent to which the creation and development of Europol can be explained from a neofunctionalist perspective.

6.5   Explaining the Development of Europol from a Nofunctionalist Perspective

There are several aspects of the initial creation of Europol and the subsequent expansion of the agency's mandate that can be explained through the neofunctionalist concept of functional spillover. The first relates to the Schengen agreement. Tömmel (2014, 180–181) argues that cooperation in the EU in Justice and Home Affairs (JHA) was initially given impetus by the main goals of the Schengen agreement: first, to abolish border controls among the states participating in the agreement; second, to produce increased cooperation in the areas of immigration, combating organised crime, and policing activities. Between 1990 and 2000, the Schengen agreement was adopted by most EU member states, as well as 3 non EU-members (Norway, Iceland and Switzerland), with the UK and Ireland choosing not to fully participate (Tömmel 2014, 181). The abolition of borders within the Schengen area (the geographical territory of participating states) can be argued to have given rise to the need for increased police cooperation among Schengen members; the freedom of movement of people increased the ease for criminals to move from country to country, and thereby increased the difficulty for the law enforcement agencies of individual states to monitor and apprehend them. The development and enhancement of police cooperation among Schengen participant states can therefore be understood as a need arising from the abolition of borders, and therefore as an example of functional spillover. From a neofunctionalist perspective, the creation of Europol as the EU's main instrument for police cooperation can be understood as the result of this functional spillover.

The second major aspect of Europol's development that can be understood in terms of functional spillover is the expansion of the agency's mandate to include a counter-terrorism role. Since the ratification of the Europol Convention, Europol has gained significantly increased powers in several areas, including counter-terrorism. Following the 9/11 terrorist attacks in New York, Europol received a greatly expanded mandate, including the creation within Europol of a Counter-Terrorist Task Force (CTTF). The CTTF is an operational centre comprised of national liaison officers from member states' police and intelligence service that provides a 24-hour service for the exchange of information related to terrorism (Rozée et al. 2014). The functional spillover that can be seen in the greatly enhanced counter-terrorism powers gained by Europol in some respects relates to the earlier spillover involving the Schengen agreement that led to Europol's creation. The 9/11 attacks led to the threat of terrorism being positioned high on the EU's security agenda; the EU's emphasis on terrorism as an EU-wide threat increased following the 2004 terrorist attacks in Madrid and 2005 attacks in London. This increased prioritisation of counter-terrorism is evident in key EU JHA official documents, including the 2004 Hague Programme and 2010 Stockholm Programme (Rozée et al. 2013). The Schengen agreement had abolished borders between participating states, and the creation of Europol had been a response to the need for increased European police cooperation to deal with the transnational movement of criminals; the heightened threat of terrorism and the transnational movement of terrorists in Europe required a similar response. Therefore, the integration that the Schengen agreement had brought to Europe had the functional spillover effect of requiring increased European police cooperation in the area of counter-terrorism. Furthermore, scholars such as Carrapico, Irrera and Tupman (2014) have discussed the connections between organised crime and terrorism, particularly in terms of the links between organised crime and the financing of terrorism. It can therefore be argued that another element of functional spillover regarding Europol's expanded mandate in counter-terrorism is that effective European police cooperation in dealing with organised crime requires effective cooperation in counter-terrorism, as the two issues are significantly connected.

Key examples of the development of Europol's powers in counter-terrorism are as follows. The CTTF was re-established after the 2004 terrorist attacks in Madrid, with an increased mandate allowing the collection of all relevant information and intelligence concerning the current terrorism threat in the EU; the analysis of the collected information and undertaking operational and strategic analysis; and the formulation of a threat assessment, including targets, modus operandi, and security consequences. The terrorist threat assessment reports have been the most important of the CTTFs new tasks, and these reports contain assessments on the financing of terrorism; information regarding terrorist movements in Europe; and the creation of an Arabic-to-English translation system used to evaluate Arabic-language intelligence (Rozée et al. 2014). In addition to the enhanced set of tasks carried out by the

CTTF, Europol has been given the competence to request that the law enforcement agencies of EU member states launch investigations as well as to share information with the United States' FBI and other third parties (Kaunert 2010, 656). Europol's AWF databases were also enhanced in 2012, reflecting the increased emphasis on counter-terrorism, with a new two-AWF structure, with one AWF containing data on counter-terrorism and the other on serious crime, and a range of focal points contained in each. This approach is intended to produce improved-quality AWFs that are in line with EU priorities (Council of the European Union 2012). Overall, these increases in Europol's powers demonstrate the on-going prioritisation of police cooperation and counter-terrorism in the Schengen area.

6.6   The Limitations of a Neofunctionalist Perspective of Europol

Despite having gained significantly increased powers since the Europol Convention was ratified in 1999, Europol has encountered several obstacles to its effectiveness that make the agency's development difficult to fully explain from a neofunctionalist perspective (Bures 2008; Rozée et al. 2013). One difficulty faced is that EU member states' law enforcement agencies have not always been willing to share information with Europol or to utilise the agency's support mechanisms (Bures 2008). If the abolition of borders resulting from the Schengen agreement necessitates the international policing cooperation that Europol facilitates (for the reasons explained above), then the reluctance of EU member states' law enforcement agencies to engage with it is problematic from a neofunctionalist perspective. While it may be the case that the removal of internal borders in the Schengen area has created the functional need for police cooperation among Schengen members, there are a number of explanations for why national law enforcement agencies have not fully engaged with Europol. First, lack of trust from member states police and intelligence agencies towards Europol has been an obstacle to cooperation. Police may be reluctant to pass sensitive and hard-earned intelligence to Europol, instead preferring to control who has access to it (Bures 2008; Interview EUR1). Furthermore, EU member states' law enforcement agencies often have well-established informal bilateral and multilateral arrangements with each other for practical coordination and sharing information. As such, they may consider these long-standing informal arrangements to be more workable, pragmatic and flexible than cooperating through Europol (Bures 2008; Interview EUR1). As well as this, according to the UK House of Lords EU Home Affairs Sub-Committee (2008), the lack of commitment from EU member states towards cooperating through Europol may, in addition to matters of trust, also be due to inadequate awareness about the functions of Europol and the value added that it can bring to policing. In addition, the fact that Europol was not formed by police professionals through a bottom up process, but rather top-down from the political and legislative bodies of the EU, may explain why member states have not always engaged fully with it (Bures 2008).

Beyond the general issues that Europol has experienced with member states reluctance to use its mechanisms, the area of counter-terrorism has been particularly difficult. There are several reasons why Europol has struggled to make progress in facilitating counter-terrorism cooperation in the EU; Bures (2008, 511) states that it is questionable whether EU member states are committed to providing Europol with a serious role in counter-terrorism, as it was mainly for political reasons that Europol's original mandate did not include counter-terrorism. Even after Europol's expanded role came to include counter-terrorism, in 2006 the EU Counter-terrorism Coordinator acknowledged that the fight against terrorism is and will remain primarily a responsibility for national authorities (Bures 2008, 512). Furthermore, as Müller-Wille argues, EU member states' law enforcement agencies have limited incentives to provide Europol with intelligence for counter-terrorism, as it is they who still have the full responsibility for providing the required intelligence support for national security. A national police service or intelligence agency could not claim that they were unable to stop a terrorist attack because Europol failed to do its job—such an explanation would not be acceptable to either the government or the public. Therefore, national services will not and are unable to rely on Europol to provide counter-terrorism intelligence and analysis (Müller-Wille 2008, in Bures 2008, 510). A final difficulty for Europol has been the lack of supranational powers provided to it under EU treaties. Despite its increased counter-terrorism mandate, Europol has played a limited role in EU counter-terrorism activities because it lacks significant operational powers (Bures, 2008; Kaunert, 2010).

The limitations of the extent to which Europol has been utilised by member states' law enforcement agencies casts some doubt on a neofunctionalist view that European integration through the Schengen agreement has necessitated (that is, had a functional spillover effect) increased EU police cooperation of the kind that Europol offers. While European expert communities in police have formed within Europol, policing experts in EU member states law enforcement agencies have been reluctant to engage fully with the agency. Furthermore, while the development of Europol's counter-terrorism mandate can be explained as a functional spillover effect related to freedom of movement and EU-level police cooperation in other areas of security, EU member states' lack of willingness to engage with Europol weakens such an analysis.

Having examined the strengths and limitations of a neofunctional analysis of European integration and the formation of expert communities in the case of Europol, the following section of this chapter will consider EU policing activities beyond the Union's borders by focusing on the case of the European Union Rule of Law Mission in Kosovo (EULEX).

6.7 The Origins of the European Union Rule of Law Mission in Kosovo (EULEX)

Over 11,000 ethnic Albanians were killed during the Kosovo War of 1998–1999, and around a million fled or were driven out of Kosovo. Most of Kosovo's Serb population left the country and were afraid to come back once the Serbian forces had left, although around 120,000–150,000 remained (Lambeth 2001, 224–230). As a response to these events, the UN Security Council adopted Resolution 1244, placing Kosovo under transitional UN administration and authorising KFOR, a NATO-led peace-keeping force. A political process then began in 2006 to determine whether Kosovo's should become an independent state or be a part of Serbia, was undertaken, based on the recommendations of a UN commissioned report that was subsequently endorsed by the UN Security Council (Eide 2005; Tannam 2013).

The European Union Rule of Law Mission in Kosovo, EULEX, is a deployment of police and civilian resources that became operational on 9 December 2008 as part of the international civil presence in Kosovo. EULEX's principal aim is to provide support and assistance to the Kosovo authorities in the area of the rule of law, particularly those aspects involving police, judiciary, and customs (Cocozzelli 2013; Kirchner 2013). Contributions to EULEX have been made by most EU member states, and by 2010 there were over 2,000 police and judicial personnel in the mission's staff. EULEX's mission statement indicates six main aims: that it will assist Kosovo authorities and law enforcement agencies in their progress toward sustainability, accountability, multi-ethnicity, freedom from political interference, and compliance with internationally recognised standards and European best practices (Council Joint Action 2008/124/CFSP, Article 2). The provision of training and Monitoring, Mentoring and Advising (MMA) to Kosovo's law enforcement agencies is the main method of introducing and maintaining the standards that EULEX seeks to achieve. While EULEX is not designed to govern in Kosovo, the mission retains executive powers in some areas; EULEX police personnel have been directly involved activities ranging from investigating organised crime and corruption, through to order maintenance tasks such as riot control. Furthermore, the mission's executive mandate includes the ability for EULEX police personnel intervene in certain cases to use "corrective powers"; for example, if the Kosovo police fail to prevent violence against non-majority communities or if the rule of law in Kosovo is being undermined by political interference (Rozée 2015). EULEX has a current mandate lasting until 14 July 2018 (EULEX 2017).

The following sections of this chapter will examine how the organisation and activities of EULEX have led to the formation of European expert policing communities and the extent to which the development of the mission can be explained from a neofunctionalist perspective.

6.8 EULEX and Expert Communities in Policing

European expert communities in policing can be seen to have formed in the case of EULEX in the mission's organisational structure and in relation to the tasks that the mission undertakes. Most EU member states have contributed resources and personnel to the mission, seconding expert police practitioners to contribute to EULEX's training of local police in Kosovo, the Monitoring, Mentoring and Advising Programme (MMA) programme, as well as the executive policing tasks that EULEX undertakes. Examples of these activities are presented below, along with an examination of how European expert communities in policing have formed and functioned in EULEX.

6.9 EULEX Training and MMA Programme

Key elements of EULEX's work in Kosovo involve reshaping local police through training and the Monitoring, Mentoring and Advising (MMA) programme; these activities are central to EULEX's mandate. Training and MMA refer to significantly different aspects of EULEX's activities. The training is provided to the Kosovo police includes a wide range of general policing tasks and generic policing duties. MMA Actions, on the other hand, set out specifically defined objectives and identify particular goals for change. EULEX's training and MMA programme both contribute to the mission's objectives in terms of setting out "an agenda for Kosovo's European perspective" (EULEX 2014), as well as to reform policing and security in Kosovo in accordance with "European best practices" and to help develop a notion of European identity in Kosovo that will generate aspiration towards future integration with the EU, ultimately leading to EU membership (European Commission DG Enlargement 2003; Tannam 2013).

The areas of training that EULEX has provided to the Kosovo police has included many areas of general police work, including day-to-day activities such as traffic policing, conducting routine patrols and ensuring public safety, through to developing principles of community based policing, building public awareness and trust of the police, as well as strengthening the concept of the police as a service, rather than just a force, among officers. A key aspect of the training that the mission has provided, and the organisational change that it has helped to produce in the Kosovo police, is the "Training the Trainers" scheme. This involves EULEX experts training senior Kosovo Police personnel, who should then pass this training on to other officers (EULEX Programme Report of 2010; Interview EEAS1).

The training offered by EULEX involves the formation of European expert communities in policing, although, as is explained below, the results of bringing such communities together have sometimes presented challenges for the mission. In addition to equipping the Kosovo police to undertake essential policing tasks unassisted, the training that EULEX provides is also intended to reform them in accordance with what the mission's mandate refers to as "best European practices and recognised international standards". In essence, this refers to reshaping the Kosovo police so that they function in a similar way to police in EU member states and carry out policing tasks according to at least the minimum standards that would be expected of the law enforcement agencies of EU member states (Interviews PR1; DGEN1; Rozée 2015). However, some difficulties have come about due to the training that EULEX provides being giving by expert police personnel from different EU member states; while EU policing experts have coordinated together in the mission to identify areas and objectives for training, the actual policing practices in different EU member states vary. This has meant that the training given to the Kosovo police has not been uniform, and has varied somewhat depending on which EU member state the EULEX trainers came from. This has not been a major obstacle for the effectiveness of the mission's training, however, and improvements have been made with regards to standardising the precise training that the Kosovo police receive (Interviews EUL1; EEAS2; Rozée 2015).

EULEX's Monitoring, Mentoring and Advising (MMA) programme differs from general training in that it identifies specific areas that where improvement is needed. Certain steps are followed by EULEX staff and local counterparts in order to produce effective change through MMA Actions; this involves a five-step process where each recommendation presented in an initial EULEX Programme Report is converted into a "Monitoring Mentoring and Advising Action", with a specific objective. This five-step process is referred to as the MMA Tracking Mechanism, which contains the following elements:

1. MMA Action Proposal – Based on assessment of performance weaknesses in the Rule of Law (first EULEX Programme Report).
2. MMA Action Specification – Based on the MMA Action Proposal and agreed by the EULEX teams and their Rule of Law counterparts.
3. MMA Action Implementation – Implementation teams prepare a detailed plan to guide implementation process and monthly reports on progress.
4. MMA Action Final Report – Final report that provides data regarding the efficacy of the MMA Action.
5. MMA Action Evaluation Report – EULEX prepares a final evaluation report on combined impact of the MMA Actions (EULEX 2014).

MMA Actions cover a wide range of areas of police work, including crime-fighting, order-maintenance and service tasks. Examples of MMA Actions include "Community Policing", which aims to build cooperative relationships between the police and the communities they serve, through to the development of "Standard Operating Procedures (SOP)" for police Operational Support Units in accordance with European best practices and internationally recognised standards, with a particular focus on Crowd and Riot Control (CRC) training (EULEX 2010; Rozée 2015).

With regards to EULEX's training and MMA programme, European expert communities in policing have been formed by police experts being seconded from many EU member states, working together to provide training and detailed MMA Actions for reshaping the Kosovo police. As noted above, differing national police practices among EU member states has let to some problems regarding the uniformity of the training provided by EULEX to local police, but progress has been seen in terms of the mission's objective to train according to "best European practices and recognised international standards" (Interviews EUL2; EEAS2).

In addition to the mission's training and MMA programme, EULEX also undertakes executive policing tasks in accordance with its mandate. These executive tasks have included a range of order-maintenance activities in response civil unrest in Kosovo as well violent attacks directed at EULEX itself. Many examples of EULEX personnel being involved in policing violent disorder have occurred in Northern Kosovo, where large numbers of Kosovo's ethnic Serb population live (Greiçevci 2011). EULEX carried out an extensive range of executive order-maintenance tasks over several months during 2009. Ethnic Serbs in Northern Kosovo had been demonstrating violently against the Albanian houses being rebuilt in Brdjani, in the northern city of Mitrovica; approximately 220,000 ethnic Serbs and other non-Albanians had been expelled from Kosovo since 1999, and riots occurred in protest due to the international community having failed to facilitate their safe return (Interview EUL1; Rozée 2015). The conflict that took place involved Serbs and Albanians throwing rocks at each other and trying to break through police cordons in the earlier stages, but escalated on 27 April 2009, when a hand grenade was thrown at EULEX police and gunshots were fired in their direction. When around a hundred ethnic Serb protestors broke through a police cordon to reach the construction site, EULEX police, with support from NATO personnel, responded to this disorder with the use of stun grenades, tear gas and rubber bullets (Reuters 2009; Security Council Report 2009; Rozée 2015).

Other instances of EULEX being involved with order-maintanance tasks in an executive capacity have occurred: EULEX police, along with NATO forces based in Kosovo, were deployed to deal with clashes that occurred on 30 May 2010 in the city of Mitrovica. At the time when Serbian local elections were being held, Ethnic Albanian protesters marched towards a bridge across the Ibar River that separates the Serbian north from the mainly Albanian regions of Kosovo. The Albanians sang nationalist songs and chanting the name of the Kosovo Liberation Army that had previously fought against the Serb forces. While this occurred, hundreds of Serb counter-protestors threw rocks at the Albanians and tried to cross the bridge towards them (Rozée 2015). After tear gas and pepper had been used to disperse the protesters, EULEX police guarded the Ibar bridge in order to ensure that no further violent disorder occurred (Interviews EUL1; EUL2; EEAS1; The Guardian 2010; Rozée, 2015) A further example of EULEX police intervening in violent clashes around the Ibar bridge occurred on 11 September 2010, following Serbia's defeat in a World Basketball Championship match; hundreds of Albanians and Serbs fought after the match, with EULEX police using tear gas to dispersing the rioting crowds (Interviews EUL2; CIV1; BBC 2010; Rozée 2015) It should be noted that EULEX's experiences of violent unrest have not only occurred in Northern Kosovo. On August 25 2009 in Pristina, the capital of Kosovo, the EULEX mission met with violent protest from ethnic Albanians against the international community's presence in Kosovo. Several police officers were injured in this protest and damage was inflicted to 28 EULEX vehicles; in this the Kosovo Police dealt with the disorder, in accordance with their training, rather than EULEX personnel carrying out executive tasks (Security Council report 2009).

With regards to the formation of European expert communities in policing, the Executive tasks that EULEX undertakes provide examples of expert police practitioners from different EU member states coordinating together to undertake "on-the-ground" policing activities. Furthermore, EULEX has the equipment and authorisation to undertake a full range of tasks related to policing public order that police have at their disposal in EU member states, from standing guard in public areas through to the use of tear gas, stun grenades and rubber bullets (Interview EUL1). Considering these executive tasks in conjunction with the many areas of policing that EULEX is involved with through training and the MMA programme, it can be seen that the mission's personnel form a European expert community that has wide-ranging involvement with policing in Kosovo.

6.10 Explaining the Development of EULEX from a Neofunctionalist Perspective

The development of EULEX can be understood through the neofunctionalist concept of functional spillover. As was discussed in the case of Europol, developments in European integration involving the Shengen area and freedom of movement have necessitated integration in European policing and security. European integration in policing focused on the internal dimension of EU security (that is, security concerns located within the EU's internal geographical area) can be seen to have produced a spillover effect resulting in integration focused on the external dimension of EU security (security concerns located or originating outside of the EU). This is because Kosovo has been considered a "gateway" for organised crime, including the movement of smuggled goods, drugs, and human trafficking to enter the EU (Rozée 2015; Mounier 2008; Hills 2009). Mounier (2008) argues that the lack of functioning security institutions is often a key reason for organised crime flourishing in post-conflict countries. Accordingly, an underlying purpose of the EULEX mission is to reform the law enforcement agencies in Kosovo in order to safeguard the EU and its member states from threats originating in the external environment (Interviews EUL1; EUL2; Rozée 2015). Therefore, part of the purpose of EULEX may be seen as to extend European integration in policing into the external dimension of security, as this is necessary for the effective functioning of policing integration in the internal dimension of security—this is an example of functional spillover.

6.11   The Limitations of a Neofunctionalist Perspective of EULEX

While the neofunctionalist explanation of the development of EULEX as the result of functional spillover from integration in EU internal policing activities is compelling, there are some limitations to such an analysis. Not all EU member states contribute to the mission, as Cyprus, Greece, Romania, Slovakia and Spain do not recognise Kosovo as an independent state (Gotev 2015; Inteview EUL1). Furthermore, the extent of member states contributions of resources to the mission differ, with some member states being reluctant to contribute valuable police personnel, preferring them to remain based in their own country (Interview EUL1; EUL2). This does not necessarily support an understanding of integration in EU external policing activities as being viewed as necessary for the success of integration in internal policing activities. Also, while EULEX's chain of command goes back to Brussels (Interview EUL1, EULEX 2017), being dependent on the willingness of some member states to contribute resources suggests that the mission is facilitated by intergovernmental cooperation, rather than being driven by supranational institutions and EU elites.

6.12   Conclusion

This chapter has considered the case of Europol and the European Union Rule of Law Mission in Kosovo (EULEX) in order to examine (a) the formation of European expert communities in policing; and (b) the extent to which a neofunctionalist perspective can explain the development of European integration in policing. In both the cases of Europol and EULEX, European expert communities were shown to have formed through expert practitioners for many EU member states being brought together to pool their expertise and achieve shared goals; e. g. Europol maintaining police analysis databases based on information obtained from all EU member states, or EULEX training local police in Kosovo in accordance with best practices and standards that exist among EU member states' law enforcement agencies. In the case of EULEX it was noted that differing policing practices among EU member states' police had meant that the training given to the Kosovo police lacked uniformity, although this issue has been identified and progress has been made in rectifying it (Interview EUL1).

A neofunctionalist analysis of the development of Europol and EULEX offered some explanation for their development as being a product of functional spillover.

In the case of Europol, the removal of national borders in the Shengen area and freedom of movement necessitated enhanced police cooperation among Shengen member states. With EULEX, the kind of "internal" European integration in policing (that is, within the EU's geographical territory) that Europol had brought necessitated integration in the "external" dimension of EU policing and security (outside of the EU), in order to prevent threats originating outside of the EU from entering it. Some limitations of neofunctionalist explanations for Europol and EULEX were noted; EU member states and their law enforcement agencies have sometimes been reluctant to utilise or contribute to Europol and EULEX; expert police practitioners' lack of commitment to EU policing mechanisms and activities does not easily correlate with the view that European integration in areas such as counter-terrorism, where Europol has received extensive additional powers, is essential for security.

In conclusion, by focusing on two of the most important and extensive examples of EU integration in policing, this chapter has shown that EU security activities have led to the formation of European expert communities in policing, and that, despite some limitations, neofunctionalism is able to offer insights into European integration in the cases of Europol and EULEX.

References

Bache, Ian, Stephen George and Simon Bulmer. 2011. *Politics in the European Union.* Oxford: Oxford Univ. Press.

BBC News. 2010. Ethnic clash in Kosovo after Serb basketball defeat. http://www.bbc.co.uk/news/world-europe-11274357.

Bigo, Didier, Sergio Carrera, Elspeth Guild and Rob Walker. 2010. *Europe's 21st century challenge: Delivering liberty and security.* London: Ashgate.

Bures, O. 2008. Europol's fledgling counterterrorism role. *Terrorism and Political Violence* 20 (4): 498–517.

Carrapico, Helena, Daniela Irrera and Bill Tupman. 2014. Transnational organised crime and terrorism: Different peas, same pod? *Global Crime* 15 (3–4): 213–218.

Cocozzelli, Fred. 2013. Between democratisation and democratic consolidation: The long path to democracy in Kosovo. *Perspectives on European Politics and Society* 14 (1): 1–19.

Council of the European Union. 2008. Council Joint Action on the European Union Rule of Law Mission in Kosovo, EULEX KOSOVO, 2008/124/CFSP. http://www.eulex-kosovo.eu/en/info/docs/JointActionEULEX_EN.pdf.

Council of the European Union. 2012. Europol work programme 2013. Brussels, 17 July, 12667/12 ENFOPOL 236. http://register.consilium.europa.eu/doc/srv?l=EN&f=ST%20 12667%202012%20INIT

Eide, Kai. 2005. A comprehensive review of the situation in Kosovo. United Nations Security Council document S/2005/635. http://www.unosek.org/docref/KaiEidereport.pdf.

EULEX. 2010. EULEX Programme Report 2010. http://www.eulex-kosovo.eu/eul/repository/docs/EPR_2010_2.pdf

EULEX. 2011. EULEX Programme Report 2011. http://www.eulex-kosovo.eu/eul/repository/docs/EPR_2011_2.pdf

EULEX. 2011. EULEX Kosovo official website: Tracking. http://www.eulex-kosovo.eu/en/tracking/. Accessed January 10, 2014.

EULEX. 2014. Official website. http://www.eulex-kosovo.eu/. Accessed January 10, 2014.

European Commission DG Enlargement. 2003. EU-Western Balkans Summit Declaration 10229/03 Presse 163 Press Release. http://europa.eu/rapid/press-release_PRES-03-163_en.htm.

Europol. 2006. AWF Factsheet. 21 June, file no. 184026. The Hague, The Netherlands.

Europol. 2016. Europol Homepage. https://www.europol.europa.eu.

Europol. 2017. About Europol. https://www.europol.europa.eu/about-europol.

Gotev, Georgi. 2015. Serbia fears EU will pressure Greece to recognise Kosovo. http://www.euractiv.com/section/enlargement/news/serbia-fears-eu-will-pressure-greece-to-recognise-kosovo/.

Greiçevci, Labinot. 2011. EU actorness in international affairs: The case of EULEX. *Perspectives on European Politics and Society* 12, (3): 283–303.

Guardian UK. 2010. Ethnic Albanians and Serbs clash in divided Kosovo town. http://www.guardian.co.uk/world/2010/may/30/kosovo-serbia-mitrovica-election-violence.

Haas, Ernst. 1958. *The uniting of Europe: Political, social, and economic forces, 1950–1957.* Stanford, CA: Stanford Univ. Press.

Hills, Alice. 2009. *Policing post-conflict cities.* New York: Zed Books.

House of Lords Select Committee on European Union. 2008. Twenty-ninth report, chapter 4: Working methods. http://www.publications.parliament.uk/pa/ld200708/ldselect/ldeucom/183/18302.htm.

Kaunert, Christian. 2010. Europol and EU counterterrorism: International security actorness in the external dimension. *Studies in Conflict & Terrorism* 33 (7): 652–671.

Kaunert, Christian, Sarah Leonard, Helena Carrapico and Stephen Rozée. 2014. The governance of justice and internal security in Scotland: Between the Scottish independence referendum and British decisions on the EU. *European Security* 23 (3): 344–363.

Kirchner, Emil. 2013. Common Security and Defence Policy peace operations in the Western Balkans: Impact and lessons learned. *European Security.* 22 (1): 36–54.

Lambeth, Benjamin. 2001. *NATO's air war for Kosovo: A strategic and operational assessment.* Santa Monica: RAND.

Lindberg, Leon. 1963. *The Political dynamics of European integration.* London: Oxford Univ. Press.

McCormick, John. 2014. *Understanding the European Union: A concise introduction.* 6th ed. New York: St. Martin's Press.

Mounier, Gregory. 2008. European police missions: From security sector reform to externalisation of internal security beyond the borders. *HUMSEC Journal* 1 (1): 47–64.

Occhipinti, John. 2003. *The politics of EU police cooperation: Towards a European FBI?* Boulder, CO: Lynne Rienner Publishers.

Reuters. 2009. EU police fire tear gas to quell Kosovo Serb protest. http://www.reuters.com/article/2009/05/01/idUSL1249216.

Rosamond, Ben. 2000. *Theories of European Integration*. New York: St. Martin's Press.
Rozée, Stephen. 2013. The European Union as a comprehensive police actor. In *European security, terrorism and intelligence: Tackling new security challenges in Europe*, ed. Christian Kaunert and Sarah Leonard, 40–64. Basingstoke: Palgrave Macmillan.
Rozée, Stephen, Christian Kaunert, and Sarah Leonard. 2013. Is Europol a comprehensive policing actor? *Perspectives on European Politics and Society* 14 (3): 372-387.
Rozée, Stephen. 2015. Order-maintenance in Kosovo: The EU as an increasingly comprehensive police actor? *European Foreign Affairs Review* 20 (1): 97–114.
Security Council Report. 2009. June 2009: Kosovo. http://www.securitycouncilreport.org/ monthly-forecast/2009-06/lookup_c_glKWLeMTIsG_b_5184949.php
Tannanm, Etain. 2013. The EU's response to the International Court of Justice's judgement on Kosovo's declaration of independence. *Europe-Asia Studies* 65 (5): 946–964.
Tömmel, Ingeborg. 2014. *The European Union: What it is and how it works*. New York: St. Martin's Press.

Research Interviews

Interview CIV1: interview with a CIVCOM official, Brussels, 30 June 2011.
Interview DGEN1: interview with an official from DG Enlargement, European Commission, Brussels, 6 July 2011.
Interview EEAS1: interview with an official from the EEAS, Brussels, 10 June 2011.
Interview EEAS2: interview with an official from the EEAS, Brussels, 18 May 2011.
Interview EUL1: interview with an official from EULEX Kosovo, Brussels, 24 June 2011.
Interview EUL2: interview with an official from the EULEX Brussels support unit, Brussels, 12 April 2011.
Interview EUR1: interview with Europol police expert, Brussels, 28 June 2011.
Interview PR1: interview with an official from a Permanent Representation to the EU, 17 June 2011.

An Epistemic-Consequentialist Social Epistemology as an Epistemological Perspective Concerning the Investigation of a Common, European Knowledge Community

7

Sebastian Nähr

Abstract

Since epistemology has traditionally been heavily individualistic in focus, the investigation of a European space of knowledge, which has its focus on a specific European space of knowledge production and on processes of knowledge exchange and dissemination in Europe, and hence on social dimensions of knowledge, at first sight does not seem to be a task of modern epistemology. However, from the 1970s on a new branch of epistemology was gradually established which tries to investigate social conditions of knowledge and cognition: Social epistemology. Based on a critical reflection of the two opposing fundamental positions in this field, revisionism and preservationism-expansionism, this chapter sketches an account of social epistemology, a so-called "epistemic-consequentialist social epistemology", which might be expected to provide a suitable framework for further epistemological investigations concerning a common European knowledge community. Following Alvin Goldman's veritistic social epistemology, this account seeks to investigate social-epistemic practices regarding their epistemic outputs while trying to overcome the shortcomings of Goldman's theory. That is to say, it presumes a classic concept of knowledge, which includes the justification condition, it stays open to an integration of epistemic relevant values other than knowledge, and it dispenses with a precise numerical quantification of a social-epistemic practice's epistemic output. Instead, an epistemic-consequentialist analysis of social-epistemic practices is based on rational considerations, aimed at plausibly qualifying a practice's epistemic output as better or worse in comparison to one of another social-epistemic practice regarding the same range of applications.

7.1 Social Epistemology between Revisionism and Preservationism or Expansionism

At first sight, the investigation of a European space of knowledge does not seem to be a task of modern epistemology. After all, epistemology, based on the philosophy of the founder of modern epistemology, René Descartes ([1641] 2009), traditionally has been heavily individualistic in focus—with some exceptions such as Thomas Reid ([1764] 1983), Charles Sanders Peirce ([1877] 1976) and Ludwig Wittgenstein ([1969] 1990), to mention but a few representatives.[1] Following a characterisation of epistemology by Michael Williams (2001), the five intertwined problem areas of epistemology, which contain the analytical problem ("What is knowledge?"), the demarcation problem ("What can be known?" and "Are there different types of knowledge?"), the problem of method ("How can knowledge be produced?"), the sceptical problem ("Is knowledge possible at all?") and the value problem ("Is knowledge associated with normativity and if so, how?"), have normally been treated from the perspective of a single and socially isolated individual. This focus meant that only individuals were seen as possible bearers of knowledge, and that only individual sources of knowledge, such as perception, or memory, were investigated in epistemology. By contrast, social conditions of cognition and knowledge have mostly been ignored in epistemology until the second half of the 20th century.[2] Such an individualistic epistemology apparently has no proper place in a project which tries to investigate something like a common, European knowledge community, focusing on a specific European space of knowledge production and on processes of knowledge exchange and dissemination in Europe and hence on social dimensions of knowledge.

However, from the 1970s on, during the so-called "social turn" in epistemology (Kitcher 1994, Scholz 2014), social conditions of cognition and knowledge, especially questions concerning knowledge production by information of others, so-called "testimony", the nature of expertise, division of cognitive labour, and groups as epistemic subjects came gradually to the fore of epistemological research (Schmitt and Scholz 2010). This social turn led to the foundation of the first scientific journal dealing with such questions in 1987 by Steve Fuller, titled "Social epistemology: a journal of knowledge, culture, and policy" and, in the same year, the publication

---

1 Current research also indicates the social dimensions of ancient and medieval epistemology. For further reading on this matter see, for instance, Hardy (2010) and Pasnau
(2010).

2 For a broader account of epistemology and its possible implications for a revisionist
social epistemology see for example Kusch (2011).

of a special issue of *Synthese* concerned with these topics (*Synthese* 73, 1987). On this basis and in the course of an intensifying philosophical discourse, the title of Fuller's journal finally became programmatic for a new branch of epistemology: Social epistemology. The term "social epistemology" itself, however, goes back to the 1950s when the librarians Margaret Elizabeth Egan and Jesse Shera (1952) proposed to establish a new research area they name "social epistemology". Its special sociological focus was intended to be on an analysis of processes of production, distribution and use of information in society (Zandonade 2004). In contrast, today, within the philosophical discipline of social epistemology, according to Oliver Scholz (2014), the following questions are typically raised: 1. What are the social conditions of individual justification and knowledge? 2. Are there social sources of justification and knowledge? 3. Can groups or institutions be bearers of beliefs, justifications and knowledge? 4. Are there experts in an objective sense? How can a layperson recognize an expert and how can she rationally judge which of two disagreeing experts to trust? 5. How should information be distributed in society? 6. How should the division of cognitive labour best be organized? 7. Which qualities of democracies do have positive epistemic influence?

These questions all highlight the social dimensions of cognition and knowledge, and thus also support the idea that the investigation of a common European knowledge community from an epistemological point of view does have a proper place within modern epistemology, namely within social epistemology. But how might such an investigation from an epistemological perspective look like? This question directly leads to a fundamental debate concerning a proper account of social epistemology, since within this new branch of epistemology two basic positions exist, the choice of which heavily influences the answer to the question above: Revisionism, and preservationism or expansionism (using the terminology introduced by Alvin Goldman [2002, 2010]). Revisionistic positions challenge fundamental concepts and projects of classic individualistic epistemology—for these positions, the social elements of social epistemology lead to epistemic-relativistic positions concerning knowledge and justification, and social epistemology is understood as the legitimate successor project of failed classic epistemology. By contrast, preservationist and expansionist positions seek to supplement the individualistic perspective on knowledge and justification of classic epistemology with the investigation of the social conditions of knowledge and justification. For these positions, the social elements of social epistemology do not lead to epistemic-relativistic positions concerning knowledge and justification, and social epistemology is understood as an extension of classic epistemology. These radically different positions in social epistemology are also reflected by the fact that, in addition to Fuller's journal, another important journal on issues of social epistemology was founded by Alvin Goldman in 2004,

called "Episteme". "Social epistemology" on the one hand understands social epistemology in a rather revisionist manner; "Episteme" on the other hand interprets
social epistemology in a more preservationist-expansionist manner.

Based on a critical reflection of these opposing positions by examining one exemplary theory from each, this chapter sketches an account of social epistemology,
a so-called "epistemic-consequentialist social epistemology", which might also work
as a framework for further epistemological investigations concerning a common
European knowledge community. To this purpose, it discusses the theory of the
probably most famous representative of a preservationist-expansionist version
of social epistemology, namely Alvin Goldman's account developed in his book
Knowledge in a social world (Goldman 1999). As a variant of a revisionist account
of social epistemology, Martin Kusch's account of a communitarian epistemology
is presented, which was developed in his *Knowledge by agreement* (Kusch 2006),[3]
since it treats classic questions of epistemology explicitly from a revisionist point
of view and, as mentioned above, it is these classic questions about the nature of
knowledge and justification which lie at the centre of the philosophical conflict
between the two different accounts of social epistemology.

7.2   Martin Kusch's Communitarian Epistemology

In his communitarian epistemology, Kusch seeks to answer the fundamental
questions of philosophical epistemology concerning the nature of knowledge and
justification. He believes that these questions of epistemology are closely interlocked
with questions of politics, especially with questions concerning the concrete forming of the organisation and production of knowledge, and he also thinks that this
interlocking means that for a correct understanding of knowledge and justification
an understanding of the social and political structure of the respective epistemic
community[4] is needed (Kusch 2006, 2). This view is a result of Kusch's epistemological considerations, which are diametrically opposed to preservationist-expansionist accounts of social epistemology: in his communitarian epistemology,

---

3   I discuss these positions in greater detail in "Grundzüge und Grundpositionen der
     Sozialen Erkenntnistheorie" (Nähr 2017).
4   The term "epistemic community" was introduced by the sociologist Burkart Holzner in
     1968 (Holzner 1968). Broadly speaking, a group of individuals with a similar perception
     of reality, who therefore share a specific form of knowledge, is called an "epistemic
     community" by Holzner. This chapter understands an epistemic community in this
     broad sense, too.

Kusch inverts the relation between the individual and community. The centre of epistemological research is no longer an individual, but an epistemic community, so that questions concerning collective epistemic players are no longer just one part of social epistemology, which itself is understood as an extension of classic epistemology. Instead, these questions form the centre of a social epistemology which treats classic individualistic questions in epistemology only as a "chapter in the epistemology of the group" (ibid., 115). This is the being the communitarian component in Kusch's position. Kusch also connects this with relativistic deliberations about meaning, truth, knowledge and justification. Thus, as a revisionist account of social epistemology, his account is a challenge for classic epistemology as well as for preservationist-expansionist accounts of social epistemology: Being feasible at least in principle, this would mean that the social turn in epistemology is a total one and classic individualistic epistemology must be replaced by a social epistemology understood in a revisionist manner.

The central statement of Kusch's epistemology is that empirical knowledge and the justification of empirical beliefs are relative to standards in an epistemic community.[5] Basically, Kusch offers two different lines of argument for this view. The first one develops an epistemic relativism (and also a relativism concerning truth) from considerations regarding the philosophy of language. Kusch advances a view called "meaning finitism", which he borrows from Barry Barnes and David Bloor (Barnes and Bloor 1982; Bloor 1991; Bloor 1997), and which he broadens for all words in a language (Kusch 2006, 159ff., 197–209, 212ff., 230–1). Although I think that Kusch is wrong about this point, this line of argument will not be treated in this paper.[6] In contrast, it is the second line of argument this chapter will be focusing on, and that is Kusch's actual communitarian account of empirical knowledge and justification. This account is a model of a dialectical theory[7], and conceives justification as well as empirical knowledge as social institutions, or as social status.

First of all, for Kusch an epistemology of empirical knowledge asks "'second-order' questions about rationality" (Kusch 2006, 86).[8] Such an epistemology presupposes the

---

5   In what follows, I will frequently use the term "justification", always to be read as a shorthand for "justification of empirical beliefs".

6   "Grundzüge und Grundpositionen der Sozialen Erkenntnistheorie" contains a brief discussion concerning this matter (Nähr 2017, 24–30). Furthermore, I do not discuss Kusch's considerations regarding testimony in this chapter, but I think that in most respects Kusch is wrong here, too (ibid., 12–24).

7   "Dialectical" refers to Aristotelian dialectics and (roughly) means dialogic negotiation (Aristoteles 1997, Topik I 1, 100a).

8   The reason for Kusch talking about "rationality" rather than "justification" is that he also includes reliably caused empirical beliefs among these second-order questions. Still,

truth of empirical beliefs and examines the justification status of empirical beliefs in general, that is, it asks what is meant by calling an empirical belief "justified". If any specific empirical belief then fulfils the criteria of justification thus obtained, this belief is an instance of empirical knowledge. On this background, Kusch's theory of justification and empirical knowledge has the following form (ibid., 150–157): An individual, empirical belief that p is justified iff the bearer of this belief forms an epistemic community with other persons, who accept the bearer's justification, so that in a sense they form and share the communal, empirical belief that p. Thereby other persons accept a justification for Kusch iff this justification sufficiently meets the norms of justification concerning the respective type of empirical beliefs in their community. This also means that the so formed communal belief will also sufficiently meet these norms. The norms are known by the members of a community because of "communally shared 'exemplars'" (ibid., 152) in the form of a pair of belief and justification, which is supposed to fulfil the norm of their type of empirical belief. It is then the logic sum of such exemplars which finally creates the norm generating and thus performative, communal, empirical belief, which is somehow fictitious and has the following form: "We believe that beliefs of type X are justified if they fulfil criteria Y; and the following are exemplary cases in which instances of X do fulfil the criteria Y: [and then follows a list of cases]" (ibid., 152ff.). This also means that the justification of an individual, empirical belief is thus accepted by other members of a community iff the pair of individual, empirical belief and justification is found to be sufficiently similar to a paradigmatic pair of individual, empirical belief and justification of the same type of empirical belief, which creates the justification norm of the respective type of empirical beliefs in their community. Hence for Kusch, justification and knowledge of an empirical belief is relative to existing norms of justification in a community, and is a social institution, a social status imposed on by members of a community. This relativity is strengthened by Kusch's opinion that a judgment of similarity between the pair of individual, empirical belief and justification in question and a paradigmatic pair of belief and justification can never be a relation of identity, so that every justification can be challenged in principle, which means that each justification is "also relative to the judgments of similarity that link a given belief-evidence pair to one or more of these exemplars" (ibid., 155). Moreover, the justification norms concerning empirical beliefs are not fixed for Kusch, but always subject to potential change—they are, in Kusch's words, "contingent" (ibid., 174)—because the exemplars, which create

---

he does not always stick to this terminology strictly (ibid., 174). By using "justification" rather than "rationality" in this chapter, reliably caused beliefs shall also be included in the concept of justification.

the norm generating communal, performative belief, might change over time and communication, so that the "communal performative belief constituting a given norm changes—more or less subtly—with each interaction" (ibid., 155).[9]

There are a couple of problems with Kusch's communitarian account of empirical knowledge and justification. Firstly, it is unclear when exactly a belief that p is justified, and when an example counts as an exemplar. How many members of the same community have to accept a justification of an empirical belief in order for it to be justified? How many members of the same community have to share an example for it to be an exemplar? These questions are probably considered to be political ones by Kusch, but the following problems directly concern Kusch's account. Following Kusch's own words, it is not feasible that an individual, empirical belief is justified and an instance of knowledge iff the bearer of the respective justification forms an epistemic community with other persons: "There is, however, a special case of beliefs to which my analysis does not apply. We might collectively agree that some genius (or some Robinson Crusoe) possesses an item of empirical knowledge the content of which is incomprehensible both to us and to any other epistemic community" (ibid., 147, n. 10). Furthermore, Kusch's considerations about norms and the acts by which they are constituted seem circular: The justification norms for empirical beliefs result from acts of justification, but these acts also depend on those norms. Bot how then was the first justification norm constituted? Kusch does not answer this question, which apparently is no problem for him, because he states that his theory does not intend to sketch a genesis of social norms (ibid., 143). That is true, but it seems that this question can just not be answered by his account at all, which does seem to be a problem for an account of empirical knowledge and justification.

Kusch's communitarian account of empirical knowledge and justification also means that no individual, empirical belief can be justified and an instance of knowledge as an individual belief independent from the acceptance of other members of the same community, and independent from the justification norms in this community. This has strong counterintuitive consequences. A belief might be as odd as possible, it still might be justified and an instance of knowledge in a suitable community. Even a person possessing pretty good evidence for the truth of an empirical belief, for example via direct observation, might then not know that p,

---

9 These remarks also clarify why empirical knowledge and justification for Kusch are connected with politics (ibid., 161ff.): They depend on norms which are valid in a community, and hence they also presuppose a community with a structural order. Without such community norms there would be no knowledge about the world either, which means that the latter is connected with knowledge about social phenomena. Finally, empirical knowledge and justification is connected with politics because the above-mentioned judgements of similarity are influenced by the social positions of the judges.

as long as the belief that p is not shared by other people of the same community and does not correspond to existing norms in that community. Kusch also states that in case that currently there is nobody who could accept a justification, we should imagine if other members of the same community would agree with our justification (ibid., 148). This, too, is not conclusive at all. Taken literally, this would mean that for each empirical belief to be justified and to be an instance of knowledge, the respective justification would have to pass a dialectic test, be it real or imagined, so that we really would have a lot to do to know the most simple facts. That there is not always a dialectic disputation needed to know something is also shown by cases of inner perception. I feel a pain in my chest, and I know from experience that I don't fancy myself having a pain in my chest, so that I think that I do have a pain in my chest. Do I now really need a dialectic disputation to have the justified belief that I have a pain in my chest? And why at all should this depend on the acceptance of any person, possibly being no doctor, and possibly being not able to competently examine my pain, and in any case not having my pain?

It surely is a crucial question concerning Kusch's account whether justification and knowledge are social institutions, or social status, as he believes. I think that he is wrong about both, but here I only want to sketch why knowledge certainly is no social institution, or social status. As Kusch himself states (ibid., 70f.): If something was a social institution created by communication, it would disappear if every talk referring to this social institution disappeared. So, would there really be no knowledge if every talk referring to the social institution of knowledge disappeared? Besides the considered cases of knowledge, which have been cases of knowledge in a form usually called "know-that", there is also the so-called "know-how", that is, the knowledge about how something is done. For example, Peter can play the piano really well and thus knows how to play the piano. Would Peter lose his knowledge of how to play the piano, and that means, would Peter not be able to play the piano anymore, just because every talk about knowledge disappeared? The term "knowledge" would disappear in that case and so would the concept of know-how, but Peter definitely would still be able to play the piano. Beyond that, Peter would still be able to teach another person how to play the piano and thereby be able to tell her such useful rules of thumb like "You don't play the piano by hitting the piano with an axe". This is interesting insofar as it clarifies the fact that it is not only a skill independent of knowledge that Peter is able to play the piano. You normally have to know, namely to know in the sense of know-that, a lot to be able to play the piano. But you do not lose this, if every talk about knowledge disappears. "Knowledge" simply is no term for social kinds, and thus knowledge is no social institution, or social status.

However, the greatest problem with Kusch's communitarian epistemology is its theoretical basis. The underlying fundament of Kusch's communitarian epistemology is a specific interpretation of the later Wittgenstein's considerations about the impossibility of a private language and about rule-following. As Wittgenstein points out in §202 of his *Philosophical Investigations* ([1953] 1984, 345), which is the crucial paragraph of his comments on rule-following and which already anticipates his so-called "private-language argument" (ibid., 356–380)[10], you cannot privately follow a rule, because otherwise to follow a rule is not distinguishable from believing to follow a rule. Analogous to the paragraphs of his private-language argument, for Wittgenstein to follow a rule presumes external criteria, which he specifies in §199: "Einer Regel folgen, eine Mitteilung machen, einen Befehl geben, eine Schachpartie spielen sind *Gepflogenheiten* (Gebräuche, Institutionen)" (ibid., 344 [emphasis in original]). For Kusch, "Gepflogenheiten", "Gebräuche" and "Institutionen" clearly are social concepts, so that Kusch thinks that Wittgenstein's important insight is that to follow a rule implies the existence of a community (Kusch 2006, 97–8): From the perspective of an individual rule-follower, only a community as an external factor can decide on standards of correct rule-following and thus enable individual rule-following at all. For Kusch, these remarks are also meaningful for normative phenomena in general, because all normative phenomena presume the difference between "is right" and "seems right", so that "normative phenomena—rules, norms, conventions, prescriptions, and standards of correctness—can exist only within communities" (ibid., 175).

However, Kusch's communitarian epistemology also harks back to an extension of these considerations based on the later Wittgenstein, namely to deliberations concerning the detailed relation between individual and community for the possibility of rule-following. Kusch advances, broadening a classification proposed by to Colin McGinn (1984), a view called "Strongest Present-Tense Community Thesis" concerning rule-following which is the following: "An individual is able to follow a rule only if the individual is currently a participating member of a group in which the very same rule is followed by other members" (Kusch 2006, 181). Although Kusch keeps it a bit in the dark, it is this thesis which is the fundament of his epistemology,[11] since the formulation of this thesis for a justification norm

---

10 In a broad sense, the private-language argument is outlined in §243 to §315. The core of the argument, however, is contained in §256 to §271.

11 Kusch is clear on the importance of this thesis: "It is this thesis that ultimately supports the communitarian epistemology proposed in this book" (ibid., 196). Unfortunately, he does not really tell us why.

of empirical beliefs (roughly) leads to the following thesis, which is necessary for Kusch's communitarian epistemology:

> An individual, empirical belief is justified iff the bearer of this belief is a participating member of a group, in which the very same justification norm for this belief is shared by other members. (SPCT-JIEB)[12]

Now, SPCT-JIEB is necessary for Kusch's account of empirical knowledge and justification, because otherwise the justification status of an individual, empirical belief does not have to depend on the actual acceptance of other members of the bearer's community, that means does not have to depend on actual shared communal norms, and thus these norms do neither have to be relative to the community nor contingent in Kusch's strong sense.

Regardless of the question whether the later Wittgenstein's considerations have to be understood in the sense of the Strongest Present-Tense Community Thesis, it is thus crucial for the plausibility of Kusch's communitarian epistemology whether this thesis is feasible or not. Kusch argues in favour of this thesis by stating that by application of a rule, followed only by an individual, it is the individual who decides which rule to follow, which means that the individual is fixing the rule's content by applying the rule, so that she cannot make a difference between following the rule correctly or incorrectly any more, because in this case every following of the rule is correct (ibid., 194f.). This for example means for Kusch that a single individual can not solve the game of Rubik's cube if nobody else is also interested in Rubik's cubes, because in this case the content of the rule for the right solution changes with circumstances and the individual's decisions, so that the individual can not make a difference between a correct and an incorrect solution of the Rubik's cube (ibid., 195f.). But what does Kusch mean by saying that the rule's content is fixed by applications of the rule? For Kusch, the common differentiation between the introduction of a rule and its application is misleading, because for him a rule only gets a *clear* content by its application (ibid., 178). He illustrates this point as follows:

> Assume my wife and I introduce the rule that we go running in the morning. Surely, when first introducing the rule, we have not foreseen all the varied circumstances that might lead us to modify the rule: visits of relatives, illnesses [...] hangovers, etc. And yet it seems natural to say that it is only in the process of our facing these varying circumstances of application that the rule itself acquires more and more content. (ibid., 178f.)

---

12 This stands for "Strongest Present-Tense Community Thesis concerning the justification of individual, empirical beliefs".

Since for Kusch there is no differentiation between the introduction of a rule and its application, because every rule's content is never clear enough for the circumstances of the rule's applications, the rule's content is fixed by applying the rule.

These considerations are not plausible at all. Firstly, it is not comprehensible why every rule should be too vague in Kusch's sense. Does this hold for all rules of artificial languages, too? What about introduction-rules for logical operators? Are they really too vague in Kusch's sense, and are they only fixed by applying them? This probably would be an astonishing result. Furthermore, the differentiation between the introduction of a rule and its application is sufficiently clear, even if the rule might be somewhat vague. In Kusch's example, the rule first was "Martin Kusch and his wife go running every morning" (rule 1). Due to certain circumstances, for example a hangover, the content of the rule changes insofar as the rule then is "Martin Kusch and his wife go running every morning, except if they suffer from a hangover" (rule 2). How can we imagine the change of the rule's content? Probably somehow like this: Martin Kusch and his wife go running every morning. Then a morning they realize that they really like running together in the morning, but only if they do not have a hangover. Rule 1 thus was too vague in the sense that this rule demanded from them to go running in the morning in cases in which both think that there are good reasons not to go running. Therefore, both finally adjust the rule, and rule 1 becomes rule 2. This adjusting surely is understandable, but the latter rule is of course different from the former one. Rule 1 and rule 2 are different rules. A change of rule's 1 content due to certain circumstances is nothing else but the introduction of a new rule—that's a sufficiently clear differentiation between an introduction and an application of a rule. Albeit from now on it is rule 2 which is in force for both, and thus rule 1 is not violated any more if they do not go running on a morning when both have a hangover, Martin Kusch and his wife are clear on the fact that they would violate rule 1 by not going running in those circumstances, if this rule still were in force—since after all they changed rule 1 exactly because this rule is violated in cases that reveal the rule to be untenable for them. But this means that Martin Kusch and his wife still can make a difference between following a rule correctly or incorrectly: Only because Martin Kusch and his wife decide themselves which rule to follow, they do not lose the possibility to differentiate between correct or incorrect rule-following. This differentiation, in fact, is the requirement for them to be able to introduce a new rule at all.

The differentiation between correct and incorrect rule-following thereby does presume a community, but in a much weaker sense than Kusch postulates. To be able to make this differentiation, an individual must only have acquired the respective knowledge in a speech community, that means she just has to know what it means to follow a rule or not to follow a rule. So, the considerations on Mr. and Mrs. Kusch made above also count for cases of individual rule-following of a rule set by oneself. Let us assume that I have set the rule not to eat chocolate after 8 pm myself. But then I once eat chocolate after 8 pm, because reading a specific book was so stressful that I needed food for my nerves and thought that it is adequate to eat chocolate after 8 pm under these circumstances. I have thus adjusted my rule, so that enjoying chocolate is no rule violation in these circumstances. Insofar, Kusch's claim that no clear differentiation between introducing and applying a rule is possible is refuted, and the fact is established that in the example I am adjusting a rule in the sense of introducing a new rule, there are no grounds left to state that in the example I do not know the difference between correct and incorrect rule-following. The rule not to eat chocolate after 8 pm is indeed not in force for me any more, but this is so because I know that if this rule were in force for me, this rule would be violated by me—therefore I have adjusted my rule after all. Finally, this also has consequences for Kusch's case of Rubik's cube. The correct solution of a Rubik's cube consists of manipulating it so that eventually each of its sides displays one uniform colour. Let us assume that Peter, because he is a member of a community in which people have been solving Rubik's cube in that way knows this, and thus Peter also knows what it means to follow a rule. Now let us assume that Peter, as the last human playing Rubik's cube in his community, adjusts the rule for the correct solution, so that a Rubik's cube is solved if just one side of the cube shows one uniform colour. This means that Peter knows that he has changed the rule, because he has introduced a new rule concerning the correct solution of Rubik's cube. If he then finally fulfils the new solution rule, he has reached the correct solution concerning the rule set up by himself, but he also knows that he has not yet reached the correct solution regarding the former solution rule in his community. In a world, or a community, in which nobody else but Peter is interested in Rubik's cubes, Peter thus still can detect a difference between a correct and an incorrect solution of a Rubik's cube. Ultimately, this means that an individual can still follow a rule, even if the very same rule is not followed by members of the same community. Hence, Kusch's Strongest Present-Tense Community Thesis and therefore also SPCT-JIEB is not plausible, which after all means that Kusch's account of empirical knowledge and justification is not feasible at all.

7.3 Alvin Goldman's Veritistic Social Epistemology

The probably most famous representative of social epistemology, Alvin Goldman, in his main work concerning social epistemology, *Knowledge in a social world* (Goldman 1999), develops a totally different account than Kusch. For Goldman, social epistemology mainly has three areas of research (ibid., 4–5): It investigates the common social sources for individual knowledge—that is, it investigates social practices relevant for individual knowledge in general—, it investigates the production and distribution of knowledge concerning individual players in epistemic groups or epistemic systems—that is, it investigates the specific (actual and possible) social practices of knowledge-production and knowledge-distribution of an epistemic group or an epistemic system—, and it investigates collectives as epistemic players.[13] In *Knowledge in a social world*, Goldman calls the social practices of any kind to be investigated in social epistemology, "social-epistemic practices" (Goldman 1999, 5ff.). For the sake of brevity, this terminology is adopted in this chapter, too.

A sketch of Goldman's classification already clarifies that social epistemology must focus on social dimensions of knowledge which complements traditional, individualistic epistemology, and thus social epistemology is not understood in a revisionist manner. Goldman's version of such a non-revisionist social epistemology thereby is geared to the concept of truth, and is therefore called "veritistic social epistemology" (ibid., 5). Collectives as epistemic players, however, do not play a crucial role in *Knowledge in a social world*. Goldman rather concentrates on the first two of the areas of research mentioned above. He does so by investigating individual social-epistemic practices, and by investigating the specific social-epistemic practices of knowledge-production and knowledge-distribution of an epistemic group or an epistemic system under the crucial question of their positive or negative influence on knowledge in comparison to falsehood or ignorance—this is what Goldman calls their "veritistic output" (ibid., 87), respectively their "veritistic value" (ibid.). This means that Goldman's veritistic social epistemology is a normative epistemology in a strong sense: A social-epistemic practice is the better and thus recommendable, the more it contributes to generating knowledge. Social-epistemic practices, in contrast, which contribute less to generating knowledge than others are not recommendable, which means that if they are current social-epistemic practices, they should be changed, and if they are possible social-epistemic practices, they should not be adopted in this form. In *Knowledge in a social world*, Goldman discusses the individual social-epistemic practices of testimony and argumentation,

---

13 This classification is in principle also advanced in more current papers by Goldman, for example in Goldman (2011a).

the transmission of information in social networks and the official regulatory procedures concerning free speech as the specific social-epistemic practices of an epistemic group[14], and the specific social-epistemic practices of the epistemic systems of science, law, democracy and education.[15]

The reason for Goldman's view that the investigation of social-epistemic practices concerning their contribution to knowledge-production and knowledge-distribution means to investigate the practice's veritistic outputs is that he understands the meaning of "knowledge" in a weak sense, namely as true belief (ibid., 24ff.). Thus, if a social-epistemic practice produces much knowledge, it produces, based on this weak sense of "knowledge", many true beliefs. In particular, two lines of argument for such a weak sense of "knowledge" are put forward in *Knowledge in a social world* (ibid.): Firstly, Goldman takes "weak knowledge" simply as a technical term. Secondly, Goldman thinks that there is a common-sense meaning of "knowledge", which supports the weak sense. Hence, truth is understood via a simple form of the correspondence theory, the so-called "descriptive success theory (DS)" (ibid., 60), which amounts to the following: "An item X (a proposition, a sentence, a belief, etc.) is true if and only if X is descriptive successful, that is, X purports to describe reality and its content fits reality" (ibid., 59 [emphases original]). This account is not intended to provide a sophisticated theory of truth, but rather serves as a technical supposition for Goldman's social epistemology.

On this background, Goldman's approach of investigating social-epistemic practices concerning their veritistic outputs, in short: Goldman's veritistic analysis, runs as follows (ibid., 87–100): Its core is the calculation of the instrumental veritistic value of social-epistemic practices. It consists of the calculation of their promotion or impediment of the acquisition of fundamental veritistic value, which is knowledge, understood as true belief. To do this, you first have to measure, in the

---

14 For Goldman this epistemic group implicitly seems to contain all technological societies.

15 This division does not exactly correspond to Goldman's in *Knowledge in a social world*. He subsumes neither the transmission of information in social networks nor the official regulatory procedures concerning free speech under the label "social-epistemic practices in an epistemic group", but calls them, together with testimony and argumentation, "generic social-epistemic practices" (ibid., 5). Unfortunately, Goldman neither tells us what generic social-epistemic practices are, nor does he explain his subsumption. I believe that the distinction proposed here is more plausible, because in an important aspect testimony and argumentation are different from the transmission of information in social networks and from the official regulatory procedures concerning free speech: The former practices are human sources of knowledge, whereas the latter are cultural phenomena of specific societies. This also corresponds to Goldman's own sketch of the programme of social epistemology as described above, and hence I think that my proposed classification of social-epistemic practices could also be in Goldman's sense.

so-called "degree of belief (DB) scheme" (ibid., 88), frequently applied by Goldman in *Knowledge in a social world*, a subject's propositional attitude towards a belief that p, which is the subject's subjective likelihood that it is the case that p. This likelihood is stated by a number within an interval between 0 and 1, where "0" symbolises that the subject does not have any confidence in the belief's truth—that is, the subject has the lowest degree of belief concerning the belief—and "1" symbolises that the subject is totally confident of the belief's truth—that is, the subject has the highest degree of belief concerning the belief. If then in fact it is the case that p, the belief that p, weighted 1, is an instance of knowledge; the belief that p, weighted 0.5, is an instance of ignorance; and the belief that p, weighted 0, is an instance of error. This is the reason why Goldman also calls the values of the degree of belief interval "veritistic values" (ibid.) of the respective beliefs—if the belief's proposition is true, the subject's belief that p is the closer a case of knowledge, in the sense of true belief, the higher the subject's subjective likelihood concerning that p is. Having fixed the veritistic value of a subject's belief you then have to measure the change of this belief's veritistic value caused by the social-epistemic practice, the veritistic value of which you wish to calculate. An example illustrates this: Let's suppose it is the case that p and at time t(1) Peter has the belief b(1) that p, with a veritistic value vv(1) of 0.3, so he is not very confident of b(1)'s truth. After obtaining additional information, however, Peter is much more confident of p's truth, so at time t(2), Peter has the belief b(2) that p, with a veritistic value vv(2) of 0.7. Thus in case p is true, the veritistic value of Peter's belief that p increases by 0.4. Now, the information which causes Peter to change his degree of belief itself is a result of a specific practice π, which of course might be a social one, like asking an expert, or a good and credible friend. Peter applies such a social-epistemic practice to the question whether p, and then, depending on the result of this application, changes his degree of belief concerning the belief that p. If this application, as in the example above, increases the respective degree of belief, and if it is the case that p, the social-epistemic practice has positive veritistic influence—that is, in the example above, this practice has the positive veritistic value of 0.4—and thus is—concerning this application—epistemically commendable for Peter. Contrarily, if it has been the case that p, and if the application has lowered Peter's degree of belief that p, the social-epistemic practice would have had negative veritistic influence and thus would have been—concerning this application—epistemically deficient for Peter.

Though these considerations already provide a good impression of Goldman's idea of a veritistic analysis of social-epistemic practices, a couple of important supplements need to be stated. Firstly, Goldman advances the principle of bivalence concerning a subject's beliefs, that is, he supposes that insofar as a subject believes that p, she also rejects the belief that non-p, and vice versa. Secondly, he also advances a question-answering model concerning the fixation of the veritistic value of a subject's belief: "In a question-answering model, agent S's belief states [...] have value or disvalue when they are responses to a question that *interests* S (or, more generally, when other agents are interested in S's knowing the answer)" (ibid., 88 [emphasis in original]). Sometimes Goldman also says that an institution's interest should play a role in this model, too (ibid., 94ff.). However, this is important because otherwise the veritistic value of a subject's total corpus of belief states, thus a subject's credibility, would absurdly also depend on questions in which the respective subject is not interested at all. Furthermore, according to Goldman, investigating a social-epistemic practice concerning its veritistic value asks for investigating a wide range of this practice's applications (ibid., 91). Such a scope should encompass all applications which typically arise, but also all types of applications which could possibly arise. Finally, it is not only possible to measure the veritistic value of a single social-epistemic practice, which Goldman calls the "absolute mode of V-evaluation" (ibid., 92), but also to compare two social-epistemic practices concerning the same range of applications regarding their veritistic values, so that it can be judged which of these two practices is veritistically better. This is called "comparative mode of V-evaluation".

Even though Goldman's considerations may be feasible prima facie, there are at least three deficiencies. Firstly, in some cases they are lacking descriptive clarity and depth. As indicated in footnote 15 of this paper, it is not clear what Goldman means by "generic social-epistemic practices" and why the practices he investigates are instances of these. The same goes for the specific systems of social-epistemic practices. Goldman does not tell us why exactly the systems he selects are instances of these or if there are any additional instances. Probably there are further interesting instances to be investigated, like for example economy, but Goldman's remarks do not really clarify this because he does not even tell us what exactly qualifies as such a system.

A further, major problem is Goldman's quantitative concept of veritistic value. This can be clarified in some respects. It has been mentioned that a belief's veritistic value depends on interesting questions, where it was tacitly assumed that these questions are deciding questions. However, as Goldman himself notes (ibid., 98–100), w-questions pose problems for his veritistic analysis. Imagine that Peter perceives an odd noise and asks himself what caused it. Unlike decision questions, there is now a wide range of possible answers. Let us further suppose that Peter is considering possible answers A(1) to A(4) and carefully assigns to them the low subjective likelihoods of 0.05, 0.02, 0.02 and 0.01 of being true. Now Gustav, who has also heard the noise, enters the stage, but is pretty confident of his answer A(5) being right, because he assigns to it the subjective likelihood of 0.8 of being true. Unfortunately, none of these answers is true.

Following Goldman's approach, the problem now is that Peter and Gustav have exactly the same credibility concerning their answer(s) on what caused the noise—that is none—, even though Peter has been more careful and therefore believes with a high subjective likelihood of 0.9 that something or somebody has caused the noise which or who really caused it. According to Goldman, this problem is initially solved by accepting that Peter is more credible than Gustav, because his beliefs of the form "x is not the cause of y" have to be considered, too. But if persons with such beliefs get more credibility, it easily can be the case that in the end this person possesses a higher credibility than a person with a correct answer to the original w-question of which, however, she is not very confident. So Goldman notes offhand that weighting the relevance of questions concerning the possible veritistic values of their answers—or their corresponding beliefs—is needed (ibid., 99f., footnote 22). This means that in the example above the question "Who has caused the noise?", and an answer to this question—or the belief which corresponds to this answer—has more relevance than the question "Who has not caused the noise?". But Goldman does not only keep quiet about how exactly such a weighting could be done, it also needs to be pointed out that such a weighting gets even more complicated, since you also have to differentiate between types of questions according to their kind and context. Imagine: If Peter has the belief that a wild pig has caused the noise, with the low subjective probability of 0.05 for being true, and in fact no wild pig has caused the noise, Peter has the belief that no pig has caused the noise with the high veritistic value of 0.95. Sure enough, this belief should not count as much as the former one regarding Peter's credibility concerning the original w-question, because otherwise a person with the right answer, to which she assigns a low subjective likelihood for being true, could easily and puzzlingly have less credibility concerning the original w-question than Peter. But it is also a difference in credibility whether Peter has the belief, with high veritistic value, that a wild pig has not caused the noise or the belief that an elf has not caused the noise. Furthermore, there is a difference in credibility between Peter having the belief, with high credibility, that a wild pig has not caused the noise when the noise was heard in a wood compared with hearing it while travelling on the underground. If such weighting is not done, persons could generate high credibility at will, simply by having abstruse beliefs. These considerations show, on the one hand, that Goldman's analysis can get pretty complex, and on the other hand already adumbrate that the quantitative dimension of this analysis, i.e., the fact that it asks for an exact numerical fixation of a belief's veritistic value, might not be adequate. Indeed, how should the weighting of relevance be determined to be numerical accurate? Some general criteria might surely be established, like: greater weight should be attached to original w-questions than to derived questions, but the more fine-grained such criteria get, the more difficult and the less accurate the fixing of values of relevance will get, too.

Another respect in which Goldman's quantitative concept of veritistic value is not plausible concerns the exact numerical fixation of a belief's degree, which is asked for in Goldman's veritistic analysis of a social-epistemic practice. This follows the wise insight that subjects normally have gradual beliefs. So, Peter may be very confident of having put a can of lemonade into the fridge, but may only be weakly convinced that Ludwig will call him today. But to express such a graduation via numerically accurate values within an interval between 0 and 1 is fairly problematic, because it is rather impossible to decide whether Peter does believe that p with the subjective likelihood of 0.3, 0.4 or perhaps 0.356 for p being true. Goldman himself is clear on this problem: "In place of 'point' probabilities, which might be too precise to capture fuzzy belief states, one might adopt a classification scheme of confidence *intervals*" (ibid., 88, footnote 13 [emphasis in original]). However, most of a subject's beliefs which do not represent instances of knowledge actually seem to be such "fuzzy belief states". But to adopt a scheme of intervals therefore means that the assignment of belief-degrees gets inaccurate relative to the extent of the interval's range.

Finally, it is also rather unclear how to measure the influence of a social-epistemic practice on a subject's subjective likelihood regarding his respective belief in a numerically accurate way. This is problematic for Goldman's solution of measuring the social-epistemic practice's veritistic values, in particular in cases of multiple social-epistemic practices. Usually a subject's degree of belief is not only influenced by one single social-epistemic practice, but by a bundle of social-epistemic practices. Imagine a judge listening to a number of witnesses' statements, followed by an expert's, say, a criminal psychologist's, testimony, which all influence his belief whether the defendant is guilty or not. How can the veritistic value of the expert's testimony be measured? Goldman advances a simple mathematical solution for this case (ibid., 97), which consists in the absolute mode of V-evaluation as well as in the comparative mode of V-evaluation in computing via counting up the single veritistic values of each of the testimonies influencing the judge's degree of belief. But if the influence of a social-epistemic practice on a subject's subjective likelihood regarding his respective belief can not be measured in a numerically accurate way, this solution becomes imprecise, that is, if you adopt the mentioned interval-scheme in this case, for example, you also obtain a solution which only is accurate relative to the interval's ranges, which surely can become rather vague.

Ultimately, Goldman's account is also deficient regarding its weak concept of knowledge. Goldman, as mentioned, derives this concept from observing common sense: "people's dominant epistemic goal, I think, is to obtain true belief, plain and simple. They want to be *informed* (have true belief) rather than *misinformed* or *uninformed*" (ibid., 24 [all emphases in original]). But, contrary to Goldman's view, I do not see why this really should support the view that there is a common-sense meaning of "knowledge" which understands knowledge as true belief. Let us look at Goldman's explanatory example for such a common sense meaning:

> Suppose it is given that P is true, and we wonder whether Jane is aware of it. The only question that needs to be resolved is whether she believes P. If she does, she is aware of it; if she doesn't, she is unaware of it. [...] "Know" can be used similarly. If we wonder whether Jane knows that P, again given its truth, the only issue to be settled is whether she believes it. She knows if she does believe it, and is ignorant (does not know) if she does not believe it (ibid., 24f.)

But aren't there many examples in epistemology which show that true beliefs may be true only by accident? In the "Jane" example, too, it can be the case that Jane truly believes that p only by accident, and probably nobody would say then that she knows that p. That is, a meaning of "knowledge" as true belief would be rather unusual in this case. Yet, wouldn't such a meaning of "knowledge" be rather unusual in this case because we commonly suppose that somebody needs good grounds for her true belief being an instance of knowledge? This, however, would mean that a common-sense meaning of "knowledge" rather asks for the classic justification condition of knowledge.

But even if there is no such common sense meaning of "knowledge" which supports any weak sense of knowledge, the concept of weak knowledge still might be defended by Goldman as being a useful technical term. However, there is indeed a stronger concept of knowledge tacitly used in Goldman's veritistic social epistemology than this postulated weak sense. As just noted, for Goldman the dominant epistemic goal of persons is to obtain true belief. This might be true somehow, but there is a potentially infinite number of true beliefs, and persons do not seek to obtain them all.[16] They do so, because for a person a true belief is not epistemically valuable per se. Only if a true belief is in a broad sense relevant for a person, this person will seek to obtain it. But then, as long as Goldman derives his weak concept of knowledge from observing the common sense, knowledge should be understood as true belief in which a person is somehow interested. And in fact, in his veritistic analysis Goldman supposes such a stronger concept of knowledge

---

16 The following argumentation is very similar to Brendel's in Brendel (2009).

implicitly, because, as noted, a person's belief is only credited with a veritistic value if either the person herself, another person, or an institution is interested in the question corresponding to the respective belief.

Goldman also tacitly uses a stronger concept of knowledge in his veritistic analysis for another reason. It would be epistemically irresponsible if a social-epistemic practice were epistemically deficient just because its single application lowered the veritistic value of a person's belief; for example, if you want to know if the practice of an expert consultation is epistemically more commendable than consulting a clairvoyant. What is needed seems to be a criterion of reliability. You could then say that an expert consultation is an epistemically more commendable social-epistemic practice than consulting a clairvoyant, even if the expert consultation might influence a person's belief veritistically negatively in a single case, whereas consulting a clairvoyant does the opposite, because the application of the social epistemic-practice of consulting an expert reliably increases the veritistic value of a person's belief, whereas consulting a clairvoyant does not do that. And indeed, Goldman implements such a criterion in his veritistic analysis implicitly, because in Goldman's veritistic analysis calculating a social-epistemic practice's veritistic value just asks for investigating the veritistic outputs of a wide range of actual and possible applications of the respective practice. In this sense, as Goldman states (ibid., 91), not only the "veritistic 'frequencies'" of social-epistemic practices must be calculated, but also their "veritistic 'propensities'". In other words, it must be evaluated how reliably a social-epistemic practice has positive veritistic influence. That his veritistic analysis implicitly presumes a criterion of reliability is stated in a footnote by Goldman himself, too: "In previous writings [...] I employed the veritistic notions of reliability and power. [...] The theoretical analysis in this chapter abandons the terms "reliability" and "power," (!) but those concepts are reflected or encapsulated in the proposed veritistic measure" (ibid., 90, footnote 16). However, Goldman's weak concept of knowledge does not provide such a criterion of reliability. Using this weak concept of knowledge in the veritistic analysis would rather mean that a social-epistemic practice is also epistemically commendable if its application produced more knowledge by chance. Thus, this weak meaning of "knowledge" does not seem to be adequate for the veritistic analysis proposed in *Knowledge in a social world*.

7.4   An Epistemic-Consequentialist Social Epistemology

The critical discussion of Kusch's communitarian epistemology in this chapter shows that his account is not convincing. Its relativistic views pose a number of problems mentioned earlier, lead to counterintuitive results, and are based on implausible assumptions. Surely it is not justified to infer from this that all revisionist accounts of social epistemology definitely are not promising at all, for at least a revisionist account does not necessarily have to be based on Kusch's Strongest Present-Tense Community Thesis. But as soon as a revisionist social epistemology—like Kusch's communitarian one—interprets the term "social" in a way that epistemic-relativistic consequences follow, it may not only be subjected to counterintuitive scenarios as mentioned in this paper, it additionally may also be subjected to pertinent and conclusive arguments against an epistemic relativism in general, like those put forward by Boghossian (2006) and Searle (1995). Alvin Goldman's veritistic social epistemology, in contrast, is clearly not subject to any of these problems. The fact that it sticks to an epistemological orientation towards the concepts of objective truth and knowledge is as convincing as its general idea of an analysis of social-epistemic practices as social epistemology's task. Hence, since epistemic-relativistic tendencies are symptomatic for revisionist accounts of social epistemology, I submit that a preservationist-expansionist account of social epistemology seems to be more plausible.

Yet in some respects, Goldman's considerations are not workable either, including their lack of descriptive clarity and depth, their weak concept of knowledge, and their veritistic analysis regarding their quantitative dimension of fixing degrees of beliefs as well as of measuring the veritistic values of social-epistemic practices. Hence, for a conclusive and comprehensive account of social epistemology it seems appropriate to clarify the nature of social-epistemic practices in general as well as to investigate the nature of specific systems of social-epistemic practices and the nature of any single social-epistemic practice. Equally fundamental are questions concerning the ontological status of collectives, which Goldman leaves completely open. On this basis, and presuming a strong concept of knowledge which also reflects the justification condition, social epistemology might then commit itself to an analysis of social-epistemic practices, which firstly must contain knowledge, understood in this strong sense, as its fundamental epistemic value. But secondly, this analysis should also, contrary to Goldman's view (1999, 245), in which knowledge is the fundamental and only epistemic value to be considered, stay open to an integration of epistemically relevant values other than knowledge. This is important for the following reason: If you analyse social-epistemic practices of science, for instance, it is important to acknowledge that the pursuit of knowledge is not the only epistemic target of this system. There are a number of additional epistemic predicates, like "being fruitful" regarding theories, which are relevant, too. In this sense, it might also be useful to investigate a social-epistemic practice's influence on the fecundity of a theory from a social-epistemic point of view. This means that such an account of social epistemology which seeks to investigate social-epistemic practices seeks to investigate the epistemic output of a social-epistemic practice in general, and has to specify initially to which epistemic value attention has to be paid. The term "epistemic-consequentialist social epistemology" therefore seems to be adequate. Nevertheless, the most important epistemic value is knowledge, so that mostly it will be the influence of a social-epistemic practice on knowledge acquisition which is to be investigated. Lastly, this analysis should dispense with a precise numerical quantification of a social-epistemic practice's epistemic output. Such an epistemic-consequentialist analysis of a social-epistemic practice's epistemic output is rather based on rational considerations, which are aimed at plausibly qualifying a practice's epistemic output as better or worse in comparison to another regarding the same range of applications. That this is possible and how (roughly) this could be done is surprisingly shown by Goldman's considerations on the veritistic commendability of different types of official regulatory procedures concerning free speech (ibid., 189–217) and by his considerations on the veritistic commendability of an author's practices regarding testimony (ibid., 105–109). Sketching parts of the latter illustrates this: A crucial question regarding the veritistic output of a testimonial speech act is whether the transmitted information is correct. Mainly, an incorrect information might be transmitted, if the author deceives herself with regard to its truth or if the author is dishonest. To minimise the frequency of dishonestly delivered testimonial speech acts, and thus to increase the veritistic output of testimony, a system of rewards concerning honesty and of retributions concerning dishonesty could therefore be established. Implementing such a system, which for example might also work via unofficial mechanisms like social proscription and social appreciation, is veritistically more commendable than not implementing such a system. Furthermore, the veritistic output of a testimonial speech act is greater the more it increases the veritistic value of a receiver's belief. That is particularly promising if the veritistic value of the receiver's belief is low, that is, if the information to be transmitted by the testimonial speech act is notably newsworthy. And to figure out, if such an information is newsworthy, an author can, for instance, evaluate whether others ask him for it. This finally means that questioning potential informants and thus forms of communication which allow this, are veristically more commendable than forms which forbid such questioning.

Within an epistemic-consequentialist social epistemology, following a useful classification of Goldman, (Goldman 2011a, 11–20) the following areas of research then are interesting: The investigation of social-epistemic practices of individual players ("SEP-Ind"), the investigation of social-epistemic practices of collective players ("SEP-Coll") and the investigation of social-epistemic practices of social systems ("SEP-Sys"). In the space of SEP-Ind, social-epistemic practices of argumentation and testimony are to be examined. These have already been in the focus of *Knowledge in a social world*, but it should also be clarified whether additional social-epistemic practices of individual players seem worth investigating, like, for example, cooperative work or cooperative learning. The investigation of testimony also suggests the inclusion of a discussion of the so-called "expert-problem" (Who is an expert? How can a layman identify an expert? How can a layman decide which of two disagreeing experts to trust?), and the discussion of the problem of disagreement between two epistemic peers regarding the same question, the so-called "peer disagreement" (Can both disagreeing epistemic peers stick to their beliefs or should they change their beliefs somehow?).[17] Both of these problems can also play a role within SEP-Sys. However, in this area of research the first task is to seek clarification what exactly systems of social-epistemic practices are and which of them need to be investigated. These systems, or rather their social-epistemic practices, then have to be examined. Within SEP-Coll, the ontological status of collectives needs to be clarified first.[18] Yet other parts of SEP-Coll investigated in *Knowledge in a social world* are the analysis of the specific social-epistemic practices of an epistemic group, like the transmission of information in social networks, and the official regulatory procedures concerning free speech. Still other practices may need to be investigated, too. Since social-epistemic practices depend on the specific epistemic group under examination, there might also be an alliance of SEP-Sys and SEP-Coll if the systems of social-epistemic practices of a specific epistemic group are to be examined.

This finally also clarifies how it might be possible to investigate a European space of knowledge from an epistemological point of view: Firstly, Europe should be understood as an epistemic community. A focus on the European Union might seem obvious, but there might be valid grounds, say cultural or geographical, to broaden this understanding of Europe as an epistemic community. Subsequently, within this epistemic community social-epistemic practices and/or special systems

---

17 For further reading regarding the expert-problem, see the two rivalling accounts of John Hardwig (1985) and of Goldman (1999, 267–271; 2011b). For further reading regarding the problem of peer disagreement see the anthology edited by Richard Feldman and Ted Warfield (2010).

18 Usually, collectives are treated as subjects with propositional attitudes (Gilbert 1989; Schmitt 1994; Tuomela 2002; List and Pettit 2011).

of social-epistemic practices and their specific practices should be identified. These social-epistemic practises should then be investigated via an epistemic-consequentialist analysis, which firstly has to select the epistemic value regarding which the practice's analysis is to be performed. For instance, when analyzing the European Union as an epistemic community, the EU's network policy regarding net neutrality might be identified as one social-epistemic practice worth investigating regarding their influence on knowledge acquisition. Moreover, European law, the European political system and the European educational system are likely to be identified as specific systems of social-epistemic practices in this community. Regarding the educational system, one could ask, for example, whether the Bologna reform in general or some of its elements in particular are to be qualified as an epistemic improvement regarding the reform's or their respective elements' influence on knowledge acquisition. It seems evident that this is an interdisciplinary task. Social epistemology, understood in an epistemic-consequentialist fashion, has to include information from different disciplines, such as political science, sociology and cultural studies, in its analyses. Ultimately,it can thus contribute to an investigation of a European space of knowledge by qualifying the social-epistemic practices of this epistemic community as epistemically more or less favourable than others.

References

Aristoteles. 1997. Topik. In *Aristoteles, Organon*, vol. 1, ed. Hans Günther Zekl, 1–447. Hamburg: Meiner.

Barnes, Barry S. and David Bloor. 1982. Relativism, rationalism and the sociology of knowledge. In *Rationality and relativism*, ed. Martin Hollis and Steven Lukes, 21–47. Oxford: Blackwell.

Bloor, David. 1991. *Knowledge and social imagery*. Chicago: Chicago Univ. Press.

Bloor, David. 1997. *Wittgenstein, rules and institutions*. London: Routledge.

Boghossian, Paul. 2006. *Fear of knowledge – against relativism and constructivism*. Oxford: Oxford Univ. Press.

Brendel, Elke. 2009. Truth and weak knowledge in Goldman's veritistic social epistemology. In *Reliable knowledge and social epistemology: Essays on the philosophy of Alvin Goldman and replies by Goldman*, ed. Gerhard Schurz and Markus Werning, 3–17. Amsterdam: Rodopi.

Descartes, Rene. (1641) 2009. *Meditationen*, ed. Christian Wohlers. Hamburg: Meiner.

Egan, Margaret Elizabeth and Jesse Shera. 1952. Foundations of a theory of bibliography. *Library Quarterly* 22 (2): 125–137.

Feldman, Richard and Ted A. Warfield. 2010. *Disagreement*. Oxford: Oxford Univ. Press.

Gilbert, Margaret. 1989. *On social facts*. Princeton, NJ: Princeton Univ. Press.

Goldman, Alvin I. 1999. *Knowledge in a social world*. Oxford: Oxford Univ. Press.

Goldman, Alvin I. 2002. *Pathways to knowledge*. Oxford: Oxford Univ. Press.

Goldman, Alvin I. 2010. Why social epistemology is real epistemology. In *Social epistemology*, ed. Adrian Haddock, Alan Millar and Duncan Pritchard, 1–28. Oxford: Oxford Univ. Press.

Goldman, Alvin I. 2011a. A guide to social epistemology. In *Social epistemology: Essential readings*, ed. Alvin Goldman and Dennis Whitcomb, 11–37. Oxford: Oxford Univ. Press.

Goldman, Alvin I. 2011b. Experts: Which one should you trust? In *Social epistemology: Essential readings*, ed. Alvin Goldman and Dennis Whitcomb, 109–133. Oxford: Oxford Univ. Press.

Hardwig, John. 1985. Epistemic dependence. *Journal of Philosophy* 82: 335–349.

Hardy, Jörg. 2010. Seeking the truth and taking care for common goods: Plato on expertise and recognizing experts. *Episteme* 7 (1): 7–22.

Holzner, Burkart. 1968. *Reality construction in society*. Cambridge, MA: Schenkman.

Kitcher, Philip. 1994. Contrasting conceptions of social epistemology. In *Socializing epistemology: The social dimensions of knowledge*, ed. Frederick F. Schmitt, 115–129. London: Rowman & Littlefield.

Kusch, Martin. 2006. *Knowledge by agreement: The programme of communitarian epistemology*. Oxford: Oxford Univ. Press.

Kusch, Martin. 2011. Social epistemology. In *The Routledge companion to epistemology*, ed. Sven Bernecker and Duncan Pritchard, 873–884. London: Routledge.

List, Christian and Philip Pettit. 2011. *Group agency: The possibility, design, and status of corporate agents*. Oxford: Oxford Univ. Press.

Mc Ginn, Colin. 1984. *Wittgenstein on meaning: An interpretation and evaluation*. Oxford: Blackwell.

Nähr, Sebastian. 2017. *Grundzüge und Grundpositionen der Sozialen Erkenntnistheorie*. Siegen: Siegen Univ. Press.

Pasnau, Robert. 2010. Medieval social epistemology: Scientia for mere mortals. *Episteme* 7 (1): 23–41.

Peirce, Charles Sanders. (1877) 1976. Die Festlegung einer Überzeugung. In *Charles S. Peirce: Schriften zum Pragmatismus und Pragmatizismus*, vol. 1, ed. Karl-Otto Apel, 149–181. Frankfurt am Main: Suhrkamp.

Reid, Thomas. (1764) 1983. Essays on the intellectual powers of man. In *Philosophical works*, ed. William Hamilton, 215–509. Hildesheim: Olms.

Schmitt, Frederick F. and Oliver R. Scholz. 2010. Introduction: The history of social epistemology. *Episteme* 7 (1): 1–6.

Schmitt, Frederick F. 1994. The justification of group beliefs. In *Socializing epistemology: The social dimensions of knowledge*, ed. Frederick F. Schmitt, 257–287. London: Rowman & Littlefield.

Scholz, Oliver R. 2014. Soziale Erkenntnistheorie. In *Grundkurs Erkenntnistheorie*, ed. Nikola Kompa and Sebastian Schmoranzer, 259–272. Münster: Mentis.

Searle, John R. 1995. *The construction of social reality*. London: Penguin.

Tuomela, Raimo. 2002. *The philosophy of social practices: A collective acceptance view*. Cambridge: Cambridge Univ. Press.

Williams, Michael. 2001. *Problems of knowledge: A critical introduction to epistemology*. Oxford: Oxford Univ. Press.

Wittgenstein, Ludwig. (1953) 1984. Philosophische Untersuchungen. In *Werkausgabe*, vol. I, 225–580. Frankfurt am Main: Suhrkamp.

Wittgenstein, Ludwig. (1969) 1990. Über Gewißheit, ed. G. E. M. Anscombe and G. H. Wright. Frankfurt am Main: Suhrkamp.

Zandonade, Tarcisio. 2004. Social epistemology from Jesse Shera to Steve Fuller. *Library Trends* 52 (4): 810–32.

About the Authors

Patricia Bauer studied Politics and Economics at the Universities of the Saarland and Hamburg. She received her PhD from the University of the Federal Armed Forces in Hamburg. She held teaching and research positions at the University of Osnabrück, Cairo University in Egypt, European Peace University in Austria and the University of Dundee in Scotland. Her research focusses on the politics of the European Union and its external relations with special respect to security, stability and democracy.

Julie Patarin-Jossec is doctoral fellow in sociology at the Centre Emile Durkheim (University of Bordeaux). After previous research on ethnographies of the laboratory, her dissertation focuses on a comparison between the European and the Russian manned space programmes using the International Space Station since 1998, aiming to provide an understanding of the social mechanisms through which science and the industry contribute to the (in)stability of the European Union and of the Russian federal state.

Sebastian Nähr received his master's degree in philosophy and social sciences in 2016. Currently, he is working as research associate for Prof. Carl Friedrich Gethmann at FoKoS, the research facility of the University of Siegen, and is preparing his PhD thesis in the philosophy of technology. Further research interests comprise philosophy of language, epistemology and social philosophy. His bachelor thesis about Kripke's philosophy of language and his master thesis about social epistemology are published by Siegen UP.

Stephen Rozée is a Lecturer in Politics at the University of Dundee, UK, where he also completed his PhD. His research interests include the European Union as a security actor with a particular focus on policing. He also holds a Master's degree in Global Politics from the University of Southampton, UK. He has published in journals such as *European Foreign Affairs Review, Perspectives on European Politics and Society and European Security.*

Bertold Schweitzer is a Lecturer in Politics and Philosophy at the University of Dundee, Scotland. He has held teaching and research positions at the European Peace University, Austria, the American University in Cairo, Egypt, and the Universities of Siegen, Wuppertal, Osnabrück, and Braunschweig, Germany. His research interests focus on the philosophy of the natural and social sciences, in particular, natural, social and cultural evolution, and the epistemology of utilising malfunctions as a device for scientific discovery and evaluation.

Thomas Sukopp is a Senior lecturer in Philosophy at the University of Siegen, Germany. After studies in chemistry, medieval and modern history, and philosophy he obtained his PhD in philosophy. He has held posts as lecturer and assistant professor at the universities of Braunschweig, Augsburg, and Bamberg. As a visiting professor, he also lectured in Taiwan and Brazil. His areas of specialization include philosophy of chemistry, debates on contemporary naturalism in philosophy of science and epistemology, intercultural philosophy of human rights, and didactics of philosophy.

Printed in the United States
By Bookmasters